THE AGE OF LIVING MACHINES

THE AGE OF LIVING MACHINES

HOW BIOLOGY WILL BUILD THE NEXT TECHNOLOGY REVOLUTION

SUSAN HOCKFIELD

W. W. NORTON & COMPANY
Independent Publishers Since 1923
New York | London

Illustrations by somersault1824 BVBA

Printed in the United States of America
First Edition

Manufacturing by Lake Book
Book design by Chris Welch
Production manager: Anna Oler

Library of Congress Cataloging-in-Publication Data

Names: Hockfield, Susan, author
Title: The age of living machines : how biology will build the next technology revolution / Susan Hockfield.
Description: First edition. | New York : W.W. Norton & Company, [2019] | Includes bibliographical references and index.
Identifiers: LCCN 2018054089 | ISBN 9780393634747 (hardcover)
Subjects: LCSH: Bionics. | Biomedical engineering. | Self-help devices for people with disabilities. | MESH: Biotechnology | Biomedical Technology | Biology
Classification: LCC TA164.2 .H63 2019 | NLM TP 248.2 | DDC 660.6—dc23
LC record available at https://lccn.loc.gov/2018054089

W. W. Norton & Company, Inc., 500 Fifth Avenue, New York, N.Y. 10110
www.wwnorton.com

W. W. Norton & Company Ltd., 15 Carlisle Street, London W1D 3BS

1 2 3 4 5 6 7 8 9 0

TO TOM AND ELIZABETH,

FOR THEIR CONSTANT PATIENCE, WISDOM, AND LOVE.

CONTENTS

PROLOGUE

For the last couple of decades, as a dean and then provost at Yale, and then as president and now president emerita of the Massachusetts Institute of Technology (MIT), I've had the privilege of looking over the scientific horizon, and what I've seen is breathtaking. Ingenious and powerful biologically based tools are coming our way: viruses that can self-assemble into batteries, proteins that can clean water, nanoparticles that can detect and knock out cancer, prosthetic limbs that can read minds, computer systems that can increase crop yield.

These new technologies may sound like science fiction, but they are not. Many of them are already well along in their development, and each of them has emerged from the same source: a revolutionary convergence of biology and engineering. This book tells the story of that convergence—of remarkable scientific discoveries that bring two largely divergent paths together and of the pathbreaking researchers who are using this convergence to invent tools and technologies that will transform how we will live in the coming century.

We need new tools and technologies. Today's world population of around 7.6 billion is projected to rise to well over 9.5 billion by 2050. In generating the power that fuels, heats, and cools our current population, we've already pumped enough carbon dioxide into the atmosphere to change the planet's climate for centuries to come, and we're now grappling with the consequences. Temperatures and sea levels are rising, and large portions of the globe are plagued with drought, famine, and drug-resistant disease. Simply scaling up our current tools and technologies will not solve the daunting challenges that face us globally. How can we generate more abundant yet cleaner energy, produce sufficient clean water, develop more effective medicines at lower cost, enable the disabled among us, and produce more food without disrupting the world's ecological balance? We need new solutions to these problems. Without them, we are destined for troubled times.

We have overcome prospects as dire as these before. In 1798, the Reverend Thomas Robert Malthus, a British cleric, economist, and demographer, observed that population growth inevitably outpaced the growth in food production. His analysis led him to warn of only one possible outcome: widespread outbreaks of famine, war, and disease. These outbreaks, Malthus claimed, would keep population growth in check—but only by the deaths of many people. "The superior power of population," he wrote, "cannot be checked without producing misery or vice."

But Malthus got it wrong. Farmers in his day had already begun to adopt new technologies, including four-field crop rotation and applying fertilizer from new sources to their crops. These new technologies fundamentally changed the equation. They made land more productive and sent more food into the marketplace. With more food available, England's population grew even more rapidly than Malthus projected, which helped to meet the work-force demands of the industrial revolution. The technology-driven agricultural revolution of the nineteenth century contributed to launching a new age of innovation and economic growth.

We've arrived at a similar moment today. Dire problems confront us with potentially disastrous consequences. Unchecked, they spell misery and devastation for much of the planet, and we do not have in hand the means to overcome them—not yet. But as I peer over the scientific horizon, I see a future that looks surprisingly bright. Biology and engineering are converging in previously unimaginable ways, and this convergence could soon offer us solutions to some of our most significant and seemingly most intractable problems. We are about to enter an era of unprecedented innovation and prosperity, and the prospects for a better future could not be more exciting.

THE AGE OF LIVING MACHINES

1

WHERE THE FUTURE COMES FROM

At an early morning meeting of the MIT Corporation on August 26, 2004, I was elected MIT's sixteenth president. My selection for the role surprised a lot of observers. Many called out the fact that I was the first woman to hold the office—a big change from my fifteen predecessors, all of whom had been men. But others noted something perhaps even more surprising: I was a biologist. I had devoted my graduate work and scientific career to understanding the physical, chemical, and structural development of the brain—not exactly the sort of thing MIT was best known for. No life scientist had previously served as the Institute's president.

When I took office, MIT had a well-deserved reputation as one of the world's premier engineering institutions and was home to internationally renowned physics, chemistry, mathematics, and computer science departments. The university had a long-standing history—based on its founding "ideas into action" mission—of collaborating with industry to transform campus discoveries into useful and marketable technologies. MIT faculty and alumni had founded an array of companies, including

Intel, Analog Devices, Hewlett-Packard, Qualcomm, TSMC (Taiwan Semiconductor Manufacturing Company), and Bose Corporation. When people thought of MIT, these were the sorts of achievements that came to mind: revolutionary products from engineering and physics that helped make the United States a leader during the explosive development of the twentieth century's electronics and digital industries.

That's why my appointment came as a surprise. A more predictable pick would have been an engineer or a computer scientist, a physicist or a mathematician. But, in fact, ever since the end of World War II, MIT has been committed to the emerging new field of molecular biology. By the time I arrived on campus, MIT's Department of Biology had taken its place among the top programs in the world. Some of its biology faculty had won the Nobel Prize for their discoveries, and several had helped to launch some of the world's top biotech companies.

With dual strengths in engineering and biology, new kinds of collaborations began. Not long after I arrived, the dean of the School of Engineering reported to me that one-third of MIT's almost four hundred engineering faculty were using the tools of biology in their work. The Institute recognized that this convergence of disciplines could create exciting ways of transforming ideas into action in the twenty-first century. In that light, my appointment made sense: we could seize the opportunity to foster the integration of biology and engineering on campus and in the international academic and industry communities.

I had to think hard about the opportunity to lead MIT. At the time, I was the provost at Yale University, where I was helping to plan a major expansion in the sciences, medicine, and engineering—a role I enjoyed immensely. A central theme in that expansion was redesigning departments and buildings to foster cross-disciplinary work. My passion for amplifying cross-disciplinary opportunities caught the attention of MIT's presidential search committee, which recognized that this sort of convergence of disciplines offered almost boundless possibilities for the future.

Could it work? Would it work? The stakes in moving between two very different institutions were high, both for me and for MIT. But in some way I had been preparing my whole life for this new assignment. So I accepted the role and embarked on what would become a fascinating journey into new fields, new ideas, and new responsibilities.

o

I have always had an insatiable desire to understand how things work, and I have always satisfied that curiosity by taking things apart. I dissected all kinds of objects even as a young child, long before I knew I wanted to be a scientist. My curiosity drove me to reduce things down to their component parts and to learn how those parts come together to give the objects their function. Emboldened by watching my father fix seemingly anything in our home, I disassembled my mother's iron and her vacuum cleaner. I opened up my favorite watch to examine its mainspring and minute gears—only to have the unwinding mainspring explode the watch out of my hands, scattering into dozens of irretrievable parts. I took my curiosity outdoors, too: I dissected daffodils in our garden and acorns that had just put forth the first sprigs of new oak trees.

How an iron worked became apparent to me after I took one apart, but how daffodils bloomed and oak trees germinated did not. How did a daffodil's brilliantly yellow petals emerge from a green bud? Why were the petals yellow instead of red? What was it inside an acorn that suddenly prompted a sprig to grow? The mysteries of living things captivated me from the very beginning. What were their mainsprings and gears?

This childhood passion for taking things apart turned into my life's work. As I came of age as a scientist, I was fortunate to grow up in the midst of two major biological revolutions. The first, molecular biology, revealed the basic building blocks of all living organisms; the second, genomics, gave molecular biology the scale necessary to identify the genes responsible for diseases and trace them across populations and species.

The importance of these two biological revolutions is impossible to overstate. Molecular biology emerged in earnest in the late 1940s and early 1950s, when a cadre of scientists, many of them physicists by training, commandeered a set of new technologies (many of which came out of the technologies developed during World War II) to describe biological mechanisms at a new, finer-grained level of resolution. They advanced our understanding of how biology works down to the level of individual molecules, hence the name "molecular" biology. Famously, James Watson, Francis Crick, Maurice Wilkins, and Rosalind Franklin used new X-ray diffraction techniques to determine the structure of DNA. This discovery opened up vast new possibilities. Scientists could now begin to understand biology at the level of the cell's "hardware"—the DNA, RNA, and protein building blocks of all living things. In time, the new tools they developed allowed them to probe the inner workings of healthy cells and advance our understanding of what goes awry in disease. Along the way, they also created important biotechnology companies, among them Genentech, Biogen, and Amgen. These companies have developed new treatments for cancer, multiple sclerosis, and hepatitis, which have saved countless lives, created tens of thousands of jobs, and contributed significantly to our economic growth.

If molecular biology made possible the study of the hardware of cells, then genomics, the next biology revolution, made possible the study of their "software"—the code that provides the instruction set for each living organism. Genomics, powered by advances in computation, has provided a map of the human genome, along with the tools for high-resolution analysis of the sequence of DNA and RNA from any species on Earth. Advances in gene sequencing and genomic data analysis that can compare genomic information among thousands of individuals have allowed scientists to begin to unravel the complex, multifactorial genetic foundations of many diseases. They have allowed biomedicine to begin developing new treatments for patients based on each individual's unique genetic makeup and disease subtype so that we have begun to be able to target individualized therapies to individual diseases. These same tools have been used to understand plants and animals and, as we'll see in the chapters ahead, to invent new solutions to some of our most pressing industrial and societal challenges.

I studied biology as an undergraduate, but that was in the years before molecular biology and genomics had fully penetrated the field. In graduate school I decided to specialize by diving into neuroanatomy, studying the brain's circuitry and how it develops. The beauty of the brain's architecture enraptured me. Using the most advanced techniques available at the time, I examined nerve cells and their exquisitely intricate interconnections. I explored how those cells assemble themselves over the course of development into the highly regular patterns that give us the ability to see, hear, think, and dream. And I studied how early experience can permanently alter both the structure and the biochemistry of the brain. Even so, I couldn't see beyond the level of cell structures to the even more fundamental building blocks of biology—namely, the proteins and other molecules that make the machinery of the brain work. Molecular biology had not yet reached neuroscience.

Shortly after finishing my PhD, I had the great fortune to be recruited to the Cold Spring Harbor Laboratory by James Watson, one of the discoverers of the structure of DNA. There I learned how biologists in other fields were using molecular biology to show how genes direct the activities of all living organisms, plants and animals alike. Flu virus, pond scum, tulips, apple trees, butterflies, earthworms, salmon, beagle puppies, humans: molecular biologists taught us that all of these organisms rely on the same set of biological building blocks.

Well in advance of most scientists, Watson grasped that the concepts and tools of molecular biology would revolutionize the study of all living things. He understood the field's power to transform biology from an observational science to a predictive science. Under his leadership, scientists at Cold Spring Harbor advanced molecular biology to reveal the mechanisms of viruses and yeast, and then used the same technologies to understand the workings of cells taken from animals and grown in dishes. Long in advance of any available technology, Watson also foresaw the possibility that the tools of molecular biology could reveal answers to many of the mysteries of the brain.

That possibility captivated me. When I started my lab at Cold Spring Harbor, neuroscience remained among the last of the biological sciences to resist the paradigm-breaking insights of molecular biology. Against the powerful currents of mainstream neurobiology, I joined a small set of adventurous neuroscientists who embraced the tools of molecular biology and began to establish a new field: molecular neurobiology.

Revolutions, even the intellectual kind, are fraught with danger and divisive forces. Fighting to advance a new approach to neuroscience put our grants, jobs, and careers on the line. Furious arguments turned staid meetings into hotbeds of rancor. One debate at an international meeting pitted scientists who studied the human brain against those who studied insect nervous systems. The argument concerned whether anything we learned from insects could enlighten us about humans. It was, fundamentally, a debate about the molecular mechanisms of evolution. And, in truth, it was more of a shouting match than a debate, because we did not yet have a "parts list" for the nervous system that would permit a definitive comparison of a human and an insect nervous system. We neither knew its genes nor could we follow their expression over the course of development.

As a small group of renegades, our band of pioneering molecular neurobiologists gradually prevailed, and the movement we launched grew into a major force that, by bringing together classical brain research and the tools of molecular biology, has transformed neuroscience, providing previously unimaginable insights into how the brain works, along with new strategies for clinical interventions. Thanks to these and other molecular biological research breakthroughs, today we have new ways of diagnosing and treating brain diseases that were intractable only a few decades ago, among them epilepsy, neurodevelopmental disorders, stroke, and inflammatory diseases like multiple sclerosis. And, it holds out reason to hope for new insights into the many still daunting diseases, including Alzheimer's and other neurodegenerative diseases.

It was indescribably exciting to be a part of a scientific revolution that brought these different disciplines and ideas together. Living and working through it, I became a participant and a proponent of what I've come to think of as a "convergence" approach to discovery.

o

I wasn't the first surprise pick as MIT's president. Early in 1930, in the midst of the Great Depression, the university chose Karl Taylor Compton, a Princeton physicist.

In retrospect Compton's appointment seems natural, even obvious, but at the time it struck many people as a break with tradition. Compton himself later claimed it was the biggest surprise of his life. From its founding in 1865, MIT had established physics as part of its core activities, but the school's reputation rested not on scientific research but on its success in technical domains. People knew it as a place that prepared engineers to make the tools and technologies that could advance the industrial age. An MIT student could expect to be prepared to pursue a career in the chemical industry or the fledgling electronics industry.

Compton inhabited a very different universe. At Princeton, he had chaired the physics department and run the nationally renowned Palmer Physical Laboratory. He had devoted much of his attention to atomic physics, an exciting field of as yet uncertain potential that had emerged just a generation earlier. The Department of Physics at Princeton was advancing fundamental science, laying the foundation for industrial applications that others would pursue.

The early twentieth century witnessed the astonishing transformation of fundamental scientific discoveries into marketplace products. As the basic components of the atom and its forces were revealed, they found their way into the entirely new electronics industry. The path from discoveries in basic physics to applications in useful products was and remains an arduous and unpredictable one. Few universities hosted both discovery and applications (science and engineering), and only a very few companies, most famously AT&T with its Bell Laboratories, invested in both fundamental discovery and new product development.

In 1897, the great physicist J. J. Thomson identified the electron as the particle that carries a negative electric charge. He and other physicists of his generation, among them Marie and Pierre Curie, Wilhelm Roentgen, and Ernest Rutherford, laid the groundwork for modeling the elementary particles that constitute all physical matter. While each pursued a somewhat different track, together they helped identify the "parts list" of components that make up and govern the behavior of the physical world: the protons and neutrons of the atomic nucleus and their surrounding cloud of electrons.

Having assembled this list, along with a set of laws that governed the behavior of the list's particles, the physicists of the era began working with engineers. Together, they now had the power to make new things: lightbulbs, radios, televisions, telephones, and even electrical systems for homes and for entire cities. The electronics industry was born, and it began putting thousands of people to work and fueling economic growth. Today, in our digitally and computationally enabled world, we continue to enjoy the fruits of that industry and the convergence of physics and engineering that made it possible.

By 1930, MIT decided it had to step up its game by raising the quality of its science departments. Recalling the feeling of this moment later in life, one member of the physics faculty wrote, "We were awakening to a whole new world of science—science in its fundamental sense, which was almost totally missing from the Institute of that time—and to a new awareness of how this modern science might transform engineering of the future." With its eye on this future and on the new integration of physics and engineering, MIT turned to Compton and offered him its presidency.

Initially taken aback, Compton was reluctant to leave his students and responsibilities at Princeton. But in the end, he came to the same realization that I would seventy-four years later: that he had been presented with the offer of a lifetime. "The magnitude of this opportunity to help science 'make good' in engineering education," he told the *Daily Princetonian*, "creates an obligation which transcends other considerations."

o

From the start, Compton devoted himself to fostering the integration of physics and engineering at MIT. He embraced the Institute's mission and recognized that the best way of developing practical solutions to problems in engineering and science was to encourage high levels of interdisciplinary collaboration. And so, many decades before I did the same, he adopted a convergence approach to discovery and innovation.

The technological demands of World War II, sometimes called "the physicists' war," brought engineering and physics even closer together. And Compton played an important role in the process. In 1933, recognizing Compton's skills as a scientist and a leader, President Franklin Delano Roosevelt appointed him as the chair of the country's new Scientific Advisory Board, which in 1940 became the National Defense Research Committee (NDRC). As the head of the NDRC at the outset of the war, Compton helped orchestrate the development of technologies such as radar, jet propulsion, and digital computing that, along with an enormous array of other technologies, proved critical to the Allies' ultimate victory. The Radiation Lab he helped create at MIT, for example, brought together almost 3,500 scientists, engineers, linguists, economists, and others in an unprecedented collaboration that invented, designed, and built radar units that have been described as "the war-winning technology."

By the war's end, under Compton's leadership, MIT was on its way to becoming the home of one of the foremost physics departments in the world, renowned for its growing strength in fundamental science, and taking its place alongside MIT's world-class engineering departments. In giving MIT dual strengths in engineering and physics, and in carrying out his broader leadership duties for the government, Compton helped chart a course for the emergence of the United States as an industrial and economic powerhouse in the decades that followed the war.

The electronics industry took off in those decades. Transistors replaced vacuum tubes, and then silicon-based circuits replaced transistors, fostering an array of discoveries and applications that opened the gates to the computer and information industries. Although Compton understood that computers would fundamentally change many aspects of communication and national defense, he could not have foreseen how the technologies he fostered would produce the digitally enabled world we live in today. Few did. That's the nature of scientific revolutions: they unfold in powerful and unpredictable ways and unleash vast possibilities. But Compton did recognize that the convergence of physics and engineering represented the beginning of a new technological age, and he did everything he could—at MIT, as a government advisor, and as a public figure—to make sure that the United States made the most of this revolution.

For these achievements alone, Compton stands as a visionary architect of the emergence of America's technological and industrial power following World War II. But during his time at MIT, he had the remarkable foresight to see another revolution coming—namely, the convergence of biology and engineering.

Compton discussed this next convergence as early as 1936, in a lecture titled "What Physics Can Do for Biology and Medicine." In it, he presented recent advances in nuclear physics, including how a new generation of cyclotrons made it possible to incorporate radioactive labels into elements. With a radioactive label, an element could be followed as it was incorporated into a molecule and then as that molecule moved through chemical reactions and metabolic pathways of a cell or an organism. The lecture prompted a physician, Dr. Saul Hertz, to ask whether this technology could be used to understand and possibly treat thyroid disease. Hertz was chief of the Thyroid Unit at Massachusetts General Hospital (MGH) and with colleagues had studied the uptake of iodine by the thyroid gland. He asked Compton whether iodine could be made radioactive. If so, he realized, it might be possible to track iodine buildup in the thyroid. That, in turn, might make it possible to diagnose thyroid disease and, perhaps, to selectively kill diseased thyroid tissue as a therapy for hyperthyroidism and thyroid cancer.

It was a bold idea, and Compton saw its merits. He connected Hertz and the MGH endocrinology group with physics colleagues at MIT, and soon this team carried through on the idea, successfully treating a set of patients with radioactive iodine, in one of the very early examples of what we would today call "precision medicine."

Compton recognized the potential of this new convergence of biology with engineering, anticipating that it would ultimately be just as powerful and as socially and economically transformative as the convergence of physics and engineering. To educate students in the hybrid field, in 1939 he described a curriculum for Biological Engineering, and in 1942 changed the name of MIT's Department of Biology to the Department of Biology and Biological Engineering. But Compton was well ahead of his time. The biologists of his day hadn't yet developed a parts list for living things of the sort that physicists had developed for physical matter—and without that list, engineers had little to work with.

Hampered by this lack of tools, the Department of Biology and Biological Engineering could not live up to its name, and within a few years it once again became the Department of Biology.

By the early 1940s, the world's attention had turned to World War II. Physics, not biology, became the necessary science. Compton worked as an extraordinarily active scientist, administrator, and public figure during the wartime years. He headed up American efforts to study radar, synthetic rubber, fire control, and thermal radiation; he ran overseas programs for the Office of Scientific Research and Development (OSRD); he served as a scientific advisor to General MacArthur; and in 1945 he became one of eight advisors appointed to guide President Truman on the use of the atomic bomb.

After the end of the war, Compton received accolades of all kinds for his contributions to the war effort. In 1946, the Army awarded him its highest civilian honor, the Medal for Merit, for his work in "hastening the termination of hostilities," and the following year the National Academy of Sciences awarded him its Marcellus Hartley Medal, for his "eminence in the application of science to the public welfare."

These two awards, along with many others, made a similar point about Compton's achievements. By bringing physics and engineering together in new ways, and by championing the revolution that this convergence enabled, he helped not only end the war but also bring about a new age of American prosperity and possibility. Compton's vision gave us an astonishing array of new tools and technologies: not just radios, telephones, planes, TVs, radar, and computers, but also nuclear power, lasers, MRI and CT scanners, rockets, satellites, GPS devices, the Internet, and smartphones. These tools and technologies have so reshaped our world that we now have a hard time conceiving of life without them.

New digital products and the digital economy they enable

continue to reshape our world. By giving rise to Big Data, the Internet of Things, and the Industrial Internet, they have made possible new business models for retail (think Amazon), hospitality (Airbnb), and transportation (Lyft, Uber). The revolution continues apace, and if Compton were still with us today, he would surely be thrilled to see its fruits.

But he would surely be just as thrilled to know that the other revolution he foresaw—the convergence of biology and engineering—is at last getting underway.

o

When I arrived at MIT, I was amazed to learn just how far down this new road many of MIT's faculty had already traveled. MIT engineers had started to incorporate the tools of biology in their work in surprising ways. Martin Polz, an environmental engineer, was using computational genomics to search for populations of plankton that capture most of the ocean's carbon dioxide. Kristala Jones Prather, a chemical engineer, was adapting microbes to make new materials, like transportation fuels and drugs. Scott Manalis, a physicist turned biological engineer, had adapted an exquisitely sensitive measuring method he had devised to weigh individual cells and monitor their growth. And inspiring all of them was Institute Professor Robert Langer, regarded as the most prolific biological engineer in the world, with over a thousand granted or pending patents, and the founder of more than twenty-five companies.

The more I learned about the incredible projects in this new realm, not just at MIT but also in labs around the world, the more convinced I became that the convergence of biology and engineering had world-changing potential. So, I made this convergence one of the major themes of my presidency, creating resources and spaces to help make it happen as rapidly as possible.

It paid off in many ways. The biology faculty that comprised MIT's Center for Cancer Research, one of the nation's preeminent centers for fundamental biological research, joined forces with engineering colleagues and reconfigured themselves into MIT's Koch Institute for Integrative Cancer Research—an exciting mash-up of engineers, clinicians, and biologists who since 2007 have been working together to understand, diagnose, and treat cancer and other diseases in new ways. Dozens of companies have spun out of the Koch Institute, many with bioengineered products that are now in clinical trials: nanoparticles that home in on cancer cells to deliver chemotherapy directly to where it matters most; imaging technologies that allow a surgeon to more accurately spot and remove cancer cells; strategies to identify infectious agents in a small fraction of the time of current methods, so that the right drug can be prescribed fast enough to save countless lives. In similar fashion, we launched the MIT Energy Initiative, which has accelerated the development of new energy technologies, many of which use components from the parts list of biology. In its first ten years, the Energy Initiative spawned close to sixty new companies that are designing new batteries, new solar cells, and new energy-management systems.

Over the course of my career, and especially during my time at MIT, I've had the great fortune to meet many of the pioneers in this emerging arena of research, and I've seen how they have translated new lab discoveries into marketplace products, turning their ideas into action. In the chapters ahead, I'll put you on the ground and in the lab with some of these key figures, and I'll introduce you to some of the ways they're hoping to use the tools and technologies that they're developing to overcome the greatest humanitarian, medical, and environmental challenges of our time.

The work they're doing is the scientific story of this century. I have no doubt about this. A century ago, physics and engineering came together and transformed our world completely, and now biology and engineering are poised to transform our future as profoundly. This book provides a preview of that emerging future, so that you, too, can enjoy the excitement of watching it happen.

I've organized the technologies in the chapters of this book to bring you, step by step, from basic to more advanced biological concepts. The new world of biology-based technologies arises from one of the most remarkable scientific revolutions. Simply put, in 1950 we did not know the physical structure of a gene or how it gives rise to physical traits. We did not know why cancer cells divide without control or what determines the color of a corn kernel, but now we do.

Chapter 2 introduces the nucleic acids, DNA and RNA, which serve as biology's information system. Nucleic acids direct the assembly of biological structures and ensure the accurate transmission of traits from one generation to the next. Nucleic acids can be manipulated, and this chapter describes how nucleic acids of viruses have been manipulated for next-generation battery fabrication. DNA and RNA carry the instruction set for the assembly of proteins, the mini-machines responsible for many biological functions, and Chapter 3 tells the story of the discovery of one such protein, called aquaporin. Aquaporin serves as a highly specific channel for water flowing into and out of cells (in bacteria, animals, and plants) and is now being deployed in commercial water filters.

The technologies in Chapter 4 introduce one of the fastest growing areas of medicine—namely, molecular medicine—with its central premise that disease processes reflect perturbations in the normal molecular processes of our cells. Highly sensitive new technologies that recognize those perturbations make early disease detection more reliable and less expensive.

Our complex biological functions, such as breathing, digestion, and hearing, are carried out by complex tissues composed of an array of different kinds of cells gathered and organized together, with the brain the most complex tissue of all. Chapter 5 describes how the brain sends messages along nerves to move limbs and how new technologies can restore to amputees and victims of brain injury the ability to move their limbs.

Chapter 6 returns us to the sum of the parts. For every living organism, the sum of gene and protein expression is revealed in its physical traits, called its phenotype. Over at least the last ten thousand years, humankind has selected and propagated plants and animals by evaluating their phenotypes. This chapter describes new engineering tools that accelerate phenotype-based selection, promising to identify more productive and more resilient food crops in time to nourish the planet's growing population.

Each in its own very different way, the technologies I describe in these chapters are all products of the revolutionary convergence of biology and engineering that we're living through today. If I'm successful in describing these new technologies that marry engineering with biology, you will see that viruses that make batteries have much in common with proteins that purify water—and with all of the book's technologies—in drawing on advances from both fields. And I hope you'll begin to recognize this common theme in many technologies just beyond the horizon.

We need to do whatever we can to enable this convergence and to bring these boundary-crossing technologies into our lives—and bring them to us fast. To that end, in the final chapter I lay out some strategies for how we can do that as quickly and as effectively as possible.

2

CAN BIOLOGY BUILD A BETTER BATTERY?

When Angela Belcher submitted her first grant proposal in 1999, one professional reviewer declared it "insane." Belcher was just about to start her first faculty position as a junior professor of chemistry at the University of Texas at Austin, and she needed grant support to launch her research career. What she was proposing to do did sound crazy: she wanted to engineer viruses that could be used to "grow" electronic circuits and, ultimately, batteries. The virus-grown batteries Belcher had in mind would charge faster than those we use today, would be produced with almost no toxic waste, and would be partly biodegradable. What she was proposing, in effect, was a clean, cheap, and natural way to make renewable energy a practical alternative to fossil fuels. The idea, Belcher felt, had world-changing potential.

Having her idea dismissed as insane stung Belcher. "When I read the review," she told me not long ago, "I cried and cried." She was upset, but she didn't quit, and today she has a record of accomplishments that, in retrospect, makes her reviewer seem insane. In 2000, she proved the viability of her unconventional idea, publishing it in *Nature*, one of the most prestigious research journals in the world. It was her first published article as an independent investigator. In 2001, recognizing her potential, MIT hired her; in 2002, *Technology Review* named her one of the best 100 innovators in the world under the age of thirty-five; in 2004, she won a MacArthur Foundation Genius Grant; and in 2006, *Scientific American* named her Research Leader of the Year. Today she is the W. M. Keck Professor of Energy at MIT where, among other roles, she directs the Biomolecular Materials Group and is an active member of MIT's Energy Initiative, leading a team designing new ways to store electricity. She has also founded start-up companies to move her lab results into marketplace products.

I met Angie Belcher early in my tenure as president at MIT. At that point, I had a lot to learn—and I had to learn it fast. I needed to figure out how MIT fostered the development of out-of-the-box ideas and how those ideas moved with astonishing rapidity into the marketplace.

To learn as much as I could as fast as I could, I invited small groups of recently tenured faculty to monthly breakfast gatherings. To get tenure at MIT, faculty members must achieve something that no one before them has accomplished, and I was confident that those invited to breakfast each month could describe the magic mix of resources, people, and ethos that enabled each of them to manage that feat. What made MIT a great place for them, and how could we make it even better? What were the exciting new frontiers they were exploring?

Over ample servings of coffee, eggs, and pastries, I asked them to tell me what they loved most about MIT and what excited them most in their research and teaching. As the conversation moved around the table, each told a story more amazing than the last, and I found myself in a future I had not yet imagined. They told me about quantum computing moving from theory to practice; about constructing drug-delivering nanoparticles layer by layer, like tiny Willie Wonka's Everlasting Gobstoppers; and dozens of other ingenious discoveries and inventions. Listening to these faculty members, a surprising observation struck me: if I had to guess what department or school they came from based solely on what they were telling me about their discoveries and passions, it would have been a difficult challenge. Their research crossed boundaries between disciplines without any celebration or permission, and I realized that flexibility was critical to the rapid translation of new ideas from the lab to the marketplace.

Many of the young faculty in those conversations straddled sets of crazily disparate disciplines, among them Belcher, whom I immediately recognized as a kind of poster child for disciplinary mixing. In addition to her work with the Biomolecular Materials Group and the MIT Energy Initiative, Belcher has appointments in the Department of Materials Science and Engineering, the Department of Biological Engineering, and MIT's Koch Institute for Integrative Cancer Research. She told me one morning that she was trying to marry biology and engineering to create a new generation of electronics, and my eyes widened with curiosity. She explained how our energy future could look very different from how we generate, distribute, and store energy today.

She first came up with her idea for a new generation of biologically constructed electronics in the 1990s while doing her doctoral research in chemistry at the University of California, Santa Barbara. She had always been fascinated by nature's ability to invent solutions in response to challenges and opportunities in the environment. During her PhD years, she developed an obsession with abalone, large sea snails commonly found along the shores of the Pacific Ocean, and how they make their shells. The process, it turns out, involves principles of bioengineering that would open Belcher's mind to all sorts of other applications including, eventually, batteries.

Evolutionarily, the abalone had to solve a very challenging problem: how to build a lightweight shell of superior strength using only simple, readily available components. The abalone has evolved an elegant and ingenious solution. First, it brings together calcium (Ca) and carbonate (CO_3), ubiquitous ocean materials, to form calcium carbonate ($CaCO_3$), the abundant mineral compound that we commonly use as blackboard chalk. Chalk itself is a weak material that crumbles easily, but the abalone overcomes that structural weakness with a two-stage manufacturing process. To start, it arranges molecules of calcium carbonate into a highly orderly array, forming small crystals. These crystals are far stronger than chalk, but they still have only 1/3,000 the strength of an abalone shell. The abalone gives these crystals the strength of steel through a process Belcher helped discover during her graduate research: it makes and distributes small filaments of protein between the crystals, creating a type of adhesive netting that functions a bit like mortar that holds together bricks in a wall. But unlike the mortar in a brick wall, the abalone shell's mortar is slightly rubbery, making the structure able to bend rather than break. The durable yet flexible protein filaments interlace with the calcium carbonate crystals, giving the abalone shell incredible dynamic strength. The shell protects the abalone while it is alive; after the abalone dies, the shell breaks down into components that replenish the resources for the next generation of shellfish—all without contributing toxic products to the environment.

Belcher keeps a collection of abalone shells in her office, and I just can't take my eyes off them whenever I visit her there. They're beautiful. Together they look like a set of unpacked Russian dolls, ranging from adorable babies that fit well within the circle of my thumb and forefinger to shells bigger than my open hand, which were probably over ten years old. One day, as we talked about the biological processes that produce magnificent materials from the most mundane elements, I couldn't resist picking up one of the biggest shells, the size of a child's baseball glove, and running my fingers over the glassy inner surface. It shimmered in a rainbow of colors as I moved it in the light.

Abalone can live up to fifty years. No matter the size, every shell has the same shape, color, and texture: rough on the outside and glossy mother-of-pearl on the inside. Each is decorated with a graceful arc of regularly spaced holes, through which the animals "breathe." It's a marvel of biological engineering, and almost from the outset, as Belcher studied how the shell was formed, she began to wonder: If abalone DNA contains the code to make proteins that can so efficiently and effectively gather elements from the sea to create shells, might we be able to commandeer the DNA from other organisms to gather other elements to do other jobs? If so, would it not be possible—as she proposed in that first grant proposal—to get viruses to gather the elements used in semiconductors, such as gallium arsenide and silicon, to make electronics? And if she could do that, then what bigger problems might she be able to use viruses to solve? Could she use them to organize the components of batteries? Her engineering mind began to race as she considered the implications. "If abalone can make all the shells they need over millions of years without emitting toxic by-products," she told me, recalling her *Aha!* moment, "why can't humans make everything we need without polluting our environment?"

Belcher grew up loving the rocks, plants, and animals of her native Texas and then also the Pacific Ocean shells along the beach during college in Santa Barbara, California. As a chemist and materials scientist, she finds the variety of shapes and sizes into which nature arranges ambient materials infinitely intriguing. Her office shelves display an array of shells, crystals, and fossils, each of whose history she has excitedly narrated for me. Once, as she cradled a beautiful crystal in one hand and a nondescript lump of whitish rock in the other, she exclaimed, "These translucent turquoise-colored crystals have the same composition as this lump of aragonite!" Along with her fascination with what nature can do, Belcher can't stop thinking about how to make the world better for the generations to come.

You and I may not spend much time thinking about whether the molecules in the materials around us have an orderly arrangement or how they organize into the materials we pick up and use every day. But Angie Belcher does. Her graduate work made her acutely aware of the importance of how materials are both composed and arranged. She showed that the abalone shell is composed of crystals of calcium carbonate bound together with a minimal amount of a mortar that the abalone makes with specific proteins. Designing better batteries relies on finding better materials and getting those materials organized into better arrangements. But improving the composition and arrangement of materials requires some pretty elaborate engineering—and it was in thinking about this that Belcher had the *Aha!* moment that led to her first grant proposal. Rather than relying entirely on human ingenuity to redesign battery components, she began to wonder if she could create a better battery by persuading viruses to arrange materials for us.

o

To understand what Angie Belcher was up against in trying to tackle the problem of energy and, more precisely, energy storage, we need to think about our energy economy. How did our patterns of energy use arise, and why does it matter where our energy will come from in the decades ahead?

Our ancestors launched the energy economy when they first learned to control fire. Discoveries of bone and plant ashes from caves in South Africa indicate that *Homo erectus*, one of our early human ancestors, made fires about 1 million years ago. Our nearer ancestors, the Neanderthals, used fire about 400,000 years ago, and archeological evidence from Pech de l'Azé I in southwestern France suggests that they could make fires on demand at least 50,000 years ago. Early humans burned grasses, branches, and tree trunks for heat and light and to cook food. Since then, we have continued to rely on nature to capture and store the energy we consume. And, lucky for us, nature does the job exceptionally well.

Plants are great energy storehouses. They store energy through photosynthesis, the chemical process that uses energy from light to combine carbon dioxide with water. Water and carbon dioxide are abundant, elementary materials, and combinations of them give rise to most of the natural materials on Earth. All it takes to make these complex materials—everything from leaves, flowers, and tree trunks to bones, skin, and muscles—is enough energy to form new chemical bonds between molecules of carbon dioxide and water. Photosynthesis converts the energy of light into chemical bonds for carbon-based building blocks. Every chemical bond is energy-in-waiting: making a bond takes energy, and breaking it releases that energy. In the case of photosynthesis, photons from sunlight provide the source of energy to make the chemical bonds, and breaking the bonds—in a fire, for example—releases energy. Burning firewood, in essence, runs the process of photosynthesis in reverse, breaking chemical bonds to release stored solar energy as heat and light. As we'll soon see in the case of batteries, stored chemical energy can also be discharged in the form of electrons.

For millennia, humans relied on trees and brush to meet energy needs. But in recent centuries our needs have grown at an accelerating pace. In the early 1800s, the primary source of energy in the United States was wood, and Americans consumed 0.4 quadrillion (10^{15}) Btu per year. (One Btu is the amount of heat required to increase the temperature of one pound of water by one degree Fahrenheit.) In 2016, total US energy consumption was 97 quadrillion Btu, 250 times more than in the early 1800s. That means that an American today consumes roughly four times more energy than an American in 1800. To meet the demands of a growing population that consumes increasingly more energy, and to find more transportable forms of stored energy, we turned to fossil fuels: energy-dense reservoirs of oil, gas, and coal, created over vast stretches of time by the compression of dead trees, plants, and other organic matter from ancient forests, the "fossils" of plant materials.

Fossil fuels—coal, gas, and oil—have higher energy density than branches and logs, so you need a lot less of them to produce the same amount of energy. This greater energy density makes them a lot easier and cheaper to move around. But we face a problem: burning carbon-based material—logs, coal, gas—releases not only the stored energy as heat and light but also the carbon dioxide captured in the photosynthetic construction of the plants. It may be hard to imagine that the seemingly vast atmosphere of Earth can't absorb all the carbon dioxide we produce in burning our carbon-based energy sources, but it cannot. Although carbon dioxide levels have risen and fallen over the course of Earth's history, those changes have always been gradual. Today we're living through something different: the unprecedentedly rapid return of extraordinary amounts of carbon dioxide into the atmosphere.

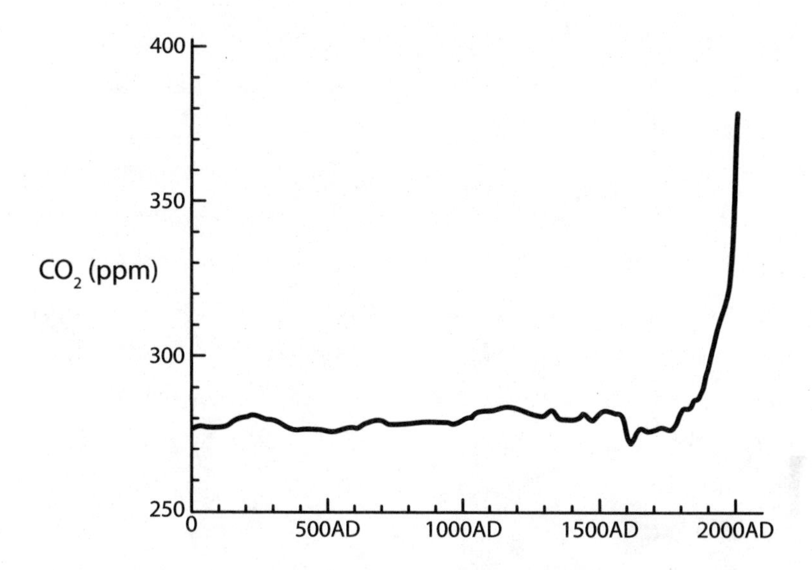

After a long period of relative stability, atmospheric carbon dioxide (CO_2; measured in parts per million [ppm]) has increased dramatically since 1800.

In a matter of only a few centuries the world population has released amounts of carbon dioxide into the atmosphere that took hundreds of millions of years to store. According to some estimates, each gallon of gasoline consumed is the product of close to 100 tons of plant material. The increase in atmospheric carbon dioxide released by our increasingly intensive burning of fossil fuels has changed the dynamics of the planet's climate and oceans dramatically, in ways that are almost certain to have dire consequences for Earth and human life.

Fossil-fuel combustion pollutes the atmosphere in other harmful ways. Standard coal burning for heating or power generation, for example, releases into the air material trapped within the coal, including mercury, sulfur, and particulates (soot), which endanger the health of people living nearby. This has serious and measurable consequences. While I understood the danger of coal burning in theory, Michael Greenstone, one of the energy economists in MIT's Energy Initiative during my time as president, brought it vividly to life for me. Greenstone told me a story from China that he and his colleagues had pieced together by meticulously studying health records. Between 1950 and 1980, Chinese national policy provided free coal for heating to people who lived north of the Huai River but not south of it. The people in coal-heated communities, they found, had a life expectancy that was 5.5 years shorter than those without government-sponsored coal—a startling difference that was accounted for almost entirely by cardiorespiratory mortality brought on by exposure to air particulates from coal burning.

If today's problems of fossil-fuel combustion were not sufficiently daunting on their own, they're about to get much worse. The world's energy demand will likely double by 2050, for two reasons. First, the world's population is on track to grow from today's roughly 7.6 billion to more than 9.5 billion over the next three decades. Second, if all goes well—and we should hope that it will—more of the world's population will be wealthy, giving more people access to the energy-intensive lifestyle that those of us in the developed world now enjoy. Today, an American consumes, on average, more than 13,000 kWh of electricity each year, while a Bangladeshi consumes, on average, only 300 kWh. What happens when more people start to consume energy at much higher levels? Are pollution and all its dangerous consequences just a necessary evil? Or can scientists and engineers invent and innovate new ways around the problem?

The world is tantalizingly rich in alternative sources of energy, in the warmth of the sun, the cooling of a summer breeze, the power of rivers and waterfalls, and the tug of the ocean's tides. I dream of a day when these alternative sources of energy might meet all our energy needs. I remember once setting out for an idyllic sail with my father and scoffing—as only an adolescent purist can—at the outboard engine he had mounted on the back of our boat. My attitude changed quickly, however, when the morning's lovely breeze died down, becalming us far from shore. Facing a day and perhaps a night without power at sea, I was suddenly very happy to burn whatever fuel it took to get us back to land, without a thought about the environment. In countless similar ways, we rely on fossil fuels whenever we need them in our daily lives—to heat our homes, to transport us and our goods all over the world, or to power our electrical grid. And most of us cannot imagine giving them up.

In recent decades, we have devised brilliant new technologies to convert sun and wind into electrical energy, and the technologies are getting better—and cheaper—all the time. But there's a hitch: although we're good at capturing and converting these sources of energy, we're not good at saving that energy for later use. The bright desert sun produces more than enough energy to warm us through the cool of the evening, and the winds that blow furiously in a storm produce more than enough energy to get us through the calm after the storm, but we haven't yet figured out how to efficiently and cost-effectively store these sources of energy. "Intermittency" has become the buzzword that describes the challenge we confront in making alternative energy sources truly practical. If scientists and engineers can manage to devise batteries that allow us to overcome the intermittency problem inherent in solar, wind, and other alternative energies, we could use these wonderfully clean and abundant sources to meet almost all of our energy needs.

Early in her career, Angie Belcher realized that she might have the tools to help solve this problem. After succeeding at what she had proposed in her early grant application, "evolving" viruses into variants that could organize nonbiological materials such as gallium arsenide and silicon for semiconductors, she began thinking about using her new tools to build batteries. Her work was timely, aligning beautifully with an emerging technology trend.

Her results with viruses and semiconductor materials gave her confidence that viruses could arrange materials at the nanoscale. She began experimenting with how far she could go in getting biological organisms to arrange nonbiological elements into useful configurations. "I wanted to know," she told me, "which of the elements in the periodic table I could persuade my viruses to use to build new structures." Not every element engaged as avidly with her viruses as others. But she found that metals and metal oxides worked particularly well. This delighted her because she recognized that these virus-binding elements could be used to make electrodes, which, in turn, seemed to open the door to a clean, efficient, and cheap new way of making batteries. But to understand how she thought she could do that, we first need to understand what batteries are.

o

Like many of the tools and technologies essential to our daily lives, batteries weren't invented to solve the problem of storing energy or to solve any practical problem at all. Batteries evolved from an insatiable desire to understand the natural world, combined with astute, curiosity-driven observation—the same process that has inspired Angie Belcher's innovations.

The first battery dates to 1800, when Alessandro Volta showed that alternating discs of copper and zinc, if stacked together and separated by a strip of salty, brine-soaked cloth, could generate an electric current. This stack has come to be known as a "voltaic pile." Put simply, it converts chemical energy to electrical energy by moving electrons from one metal to another.

Electrons are tiny, negatively charged particles. The copper disc in a voltaic pile forms the positive pole, or cathode, and the zinc disc forms the negative pole, or anode. The collection of alternating discs of a first-generation battery doesn't convey any charge on its own, but if something that conducts electricity, like a metal wire, is connected between the two poles, it will conduct the flow of electrons from the anode to the cathode. Including an electric device, like a small lightbulb, in the circuit allows the current flow to be monitored—when the current flows, the bulb lights up. Eventually, this process runs its course, when all the available electrons have transferred and the metals have exhausted their capacity to generate or accept new electrons. At that point, the battery can no longer generate electricity, the lightbulb fades to off, and the battery must be replaced.

Volta experimented with different cathode and anode metals and different electrolyte fluids to maximize electricity output, but he never produced enough electricity for any practical application. Others picked up where he left off, however, and built batteries that could power electrical devices. If he visited us today, Volta would recognize our batteries as direct descendants of his.

One way to think of standard batteries is as energy transportation devices. The standard AAA batteries that I carry with me on an airplane to listen to music on my headphones are merely a way to transport chemical energy that I can convert into electrical energy whenever I want. When I'm at home, an electrical wall outlet provides a much more abundant source of electricity, but when household electricity isn't available, a battery fills the gap. Rechargeable batteries, like the ones inside your phone or laptop, function even better as energy storage devices. When they lose their charge, they

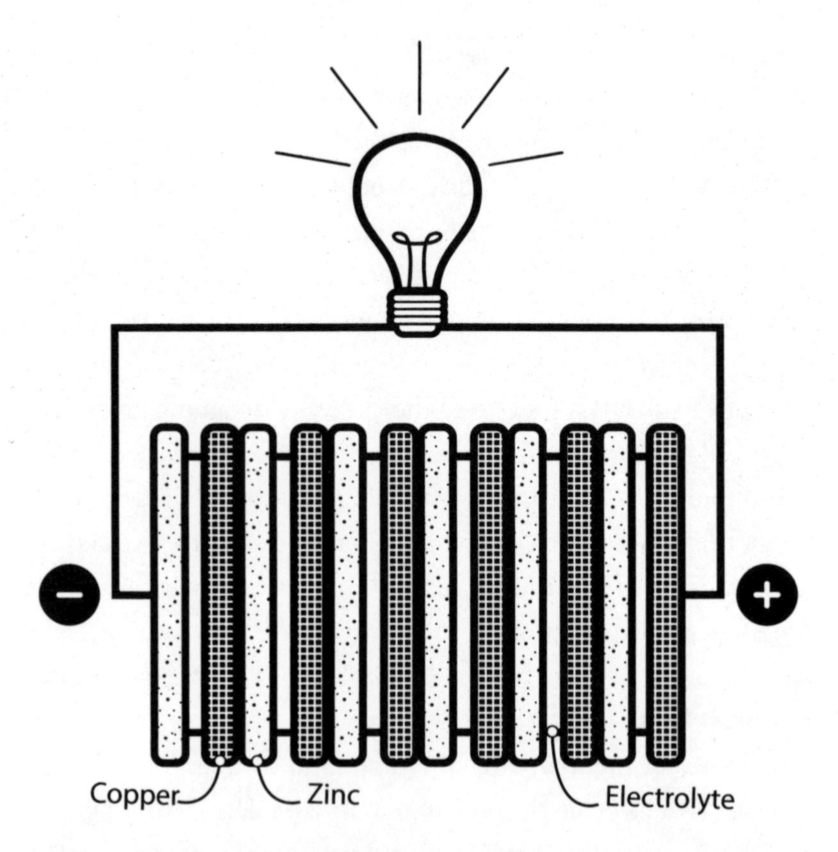

A basic voltaic battery consists of alternating pairs of copper and zinc discs with a layer of electrolyte between each pair. Electrons flow from the zinc anode (—) to the copper cathode (+), traveling through the wire and turning on a small lightbulb.

can run their discharging process in reverse, using electricity from an external source—a wall plug, say—to recover the electrons that migrated from the anode to the cathode back to their starting positions in the anode. Then another cycle of discharging can begin.

You might think this reversible process should be easy. But it's not. It requires materials that can serve as both electron generators and electron acceptors. The first successful rechargeable battery came into use just over a century ago. It used lead for the electrodes and sulfuric acid as an electrolyte. These components made the batteries heavy and dangerous, but nonetheless they had remarkable reliability, and as a result, we still use lead-acid batteries for many of our standard heavy-duty recharging needs. Most cars on the roads around the world still use lead-acid batteries. And even though most electric cars use lithium-ion batteries for driving power, most still use lead-acid batteries to power their headlights, fans, and safety features.

In more recent years, the proliferation of mobile electronic devices has been fueled by the development of vastly lighter and safer rechargeable batteries. Lithium-ion batteries now power our cell phones, flashlights, and most of our portable electronic devices, but they're not sufficiently economical, efficient, and powerful to meet large-scale energy demands. They also pose risks, with the batteries of hoverboards and cell phones sometimes catching fire. Beyond the technical limitations of the batteries themselves, standard manufacturing processes for batteries require very high temperatures (think energy consumption) and produce toxic byproducts. Estimates of how much have varied, but in 2017 the IVL Swedish Environmental Research Institute calculated that building an electric car battery might generate as much waste as the equivalent of 20 tons of carbon dioxide, which is the amount produced when 2,250 gallons of gasoline are burned. So, before you're tempted into feeling virtuous about driving your electric car, you should first factor in the battery manufacturing costs (in energy consumption and carbon dioxide production) and the source of the electricity for recharging its battery (does the electricity for recharging come from a hydroelectric plant or from a power plant that burns fossil fuels?). In addition to the energy cost of fabrication, making a battery, like many manufacturing processes, generates plenty of waste and some of it is highly toxic.

All things considered, the options available to us today cannot keep up with our rapidly growing energy-storage demands. Coming up with an alternative approach to energy storage is exactly what motivated Angie Belcher to get into the game. She isn't the only one working on developing cleaner, lighter, and more efficient batteries, of course. Researchers all over the world have taken on this problem, and in recent years dozens of promising new technologies have made their way out of research labs. Yi Cui and his colleagues at Stanford University, for example, have designed nanoparticles for batteries that pack more tightly, carry more charge, and hold the promise of lasting longer than current batteries. Ze Xiang Shen, at the Nanyang Technical University in Singapore, is developing sodium-ion batteries using nanosheet designs as a potentially low-cost and safe alternative to lithium-ion batteries. But Belcher has taken an astonishingly revolutionary approach by turning to biology—and, specifically, to viruses—for help.

o

Unlike most living organisms including plants, animals, and even single-celled yeast and bacteria, viruses lack most standard components for life. They don't have cell walls, or nuclei, or any internal structural elements that other living organisms have. Instead, they consist of little more than a protein capsule that encloses a strand of DNA or RNA. That's all! Nonetheless, they have flourished in every ecosystem of our planet for eons; we have evidence of an insect-infecting virus that dates back 300 million years. Viruses reproduce with remarkable, even alarming, success; they populate all sorts of different environments, including the human body, often causing some of our most dreaded diseases.

Viruses are biological minimalists. But even in their simplicity, a virus will look very much like its parent, much as the color of my daughter's hair and eyes matches mine. But viruses can't do a lot on their own. While they carry their own information set in their DNA or RNA, they lack the machinery to propagate themselves. To survive and propagate, viruses depend on a host organism. They infect the cells of animals and plants—and when that happens to us, we experience it as a viral disease.

Viruses use the same molecules to transmit their genetic instruction set to their offspring that we use to pass on our genetic information to our children. The information in their genes and ours is carried from generation to generation in molecules of nucleic acid, either DNA or RNA. The structure of nucleic acids gives them two essential functions: making exact replicas of themselves (critical for transmitting genetic information to subsequent generations) and directing the assembly of proteins, the building blocks of every organism.

The accurate duplication of nucleic acids is essential for precise transmission of genetic information between parents and offspring. DNA, deoxyribonucleic acid, has a ladder-like structure, with two parallel backbones interconnected by a series of rungs. Each rung consists of a pair of molecules called bases, which come in four types: adenine (A), guanine (G), cytosine (C), and thymine (T). The two bases in each rung always pair in the same way: A with T and G with C. The information content of nucleic acids is written in the order of the bases running along the ladder's backbone. RNA, ribonucleic acid, exists as only one-half of a DNA ladder and substitutes thymine (T) with uridine (U). The forced pairing of bases, G-C and A-T (or A-U), dictates accurate copying of DNA (or RNA) into the next generation of a cell or organism. During cell division, the DNA ladder splits lengthwise into two half-ladders, dividing each rung into a one-base half.

Each half-DNA strand serves as a template for a new half, directed by the forced A-T and G-C pairing.

In 1953, when James Watson and Francis Crick first reported the structure of DNA, in a one-page paper published in *Nature*, they concluded with one of history's greatest understatements, "It has not escaped our notice," they wrote, "that the specific pairing we have postulated immediately suggests a possible copying mechanism for the genetic material." This little remark proved not only to be correct but also presaged the start of a dramatic new era for biology—namely, molecular biology.

The DNA or RNA of a cell or a virus directs the assembly of proteins, the building blocks for every organism's form and function. The sequence of bases along the virus DNA (or RNA) is the code that directs the composition of the fewer than a dozen different viral proteins. Similarly, the sequence of bases in our DNA directs the composition of all of our proteins. For Angie Belcher to use viruses to make batteries, she began manipulating benign strains of lab viruses to enable them to organize the components of batteries.

Viruses replicate through a fascinating, aggressive, and evolutionarily very successful process. Without their own machinery for most biological processes, they invade cells of other organisms and parasitize their replication machinery. Proteins on the external coat of a virus cleverly bind to specific proteins on the surface of host cells. Some viruses bind to our cells: the flu virus binds to cells along our respiratory tract, whereas the hepatitis C virus binds to our liver cells. Other viruses bind to other animal and plant cells. Once attached to a host cell, a virus injects its DNA or RNA, which then takes control of the host cell's machinery, directing the host to reproduce viruses in huge numbers. This puts such a burden on the host cell that it either slows its own processes,

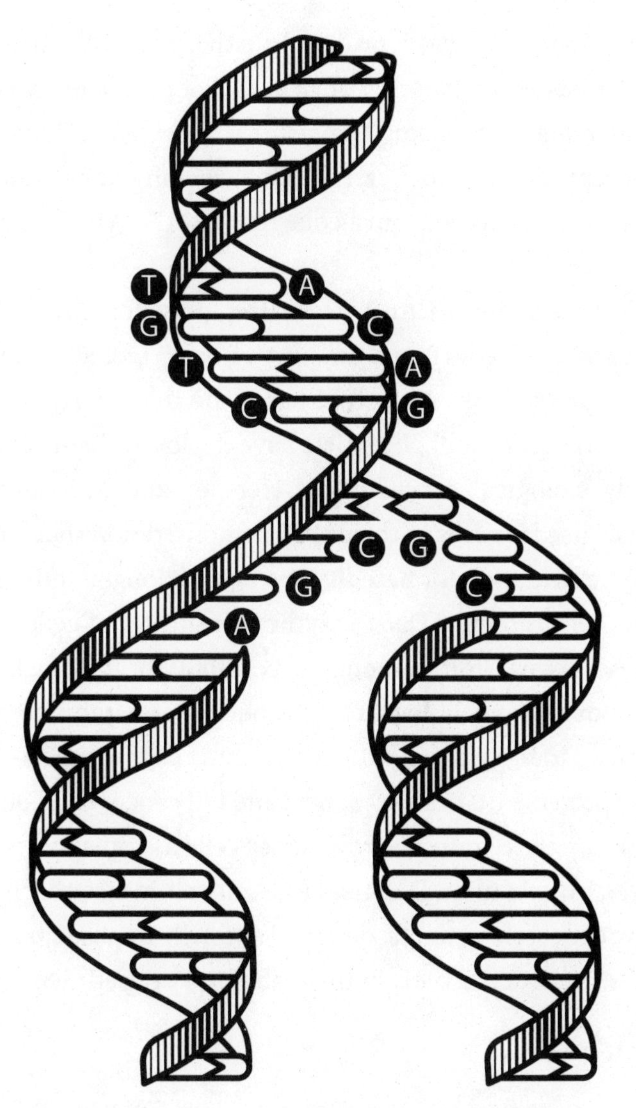

The structure of DNA has self-replicating features. At the top of the figure, the double-stranded, twisting DNA ladder has rungs, with each rung composed of a pair of bases. The bases always pair in the same manner: C with G and A with T. During DNA replication, the bases separate (at the middle of the figure). Each of the now unpaired bases will pair with an appropriate partner (A with T and C with G), resulting in the assembly of two new strands (at the bottom of the figure), each identical to the original strand.

often to a near standstill, or dies. In either case, the virus-inhabited cell releases a new army of viruses that invade other cells and reproduce inside of them. The result is an explosively rapid process of reproduction that can wreak havoc on our health, as anybody who has suffered from a cold, the flu, HIV-AIDS, or hepatitis well knows.

As troublesome a threat as viruses pose for our health, we have learned an enormous amount of basic biology from them. Their elegantly simple structure has made them particularly useful as a laboratory tool. Scientists have deployed them for decades to study biological processes. Al Hershey and Margaret Chase famously used viruses in their 1952 demonstration that DNA carries the information for heredity, resolving a long-standing debate on whether protein or DNA was the transmitter of genetic traits.

Viruses have become among the best tools for moving DNA and RNA from cell to cell. For example, new cancer immunotherapies use viruses to ferry into a patient's immune cells genes that encode specific proteins that can recognize and kill cancer cells, equipping the immune system to destroy cancer cells as if they were foreign invaders. Viruses move DNA and RNA so well, in fact, that researchers now use many variants, designed specially to pose no danger to humans, as standard tools in their laboratory experiments.

o

Viruses come in many shapes. Some are twenty-one-sided icosahedrons, some are simple spheres, and some look like tiny rocket ships with landing gear on one end. Each has, over very long evolutionary time, optimized its structure for survival. None, of course, has evolved to build a battery, but Angie Belcher recognized that the structure of one virus in particular, the M13 bacteriophage, makes it almost ideal for solving some of the problems that a battery must overcome. And she has figured out how to direct the evolution of M13 to turn it into a micro-factory for battery assembly. Her M13-based batteries can pack more power into a smaller package, and they can make batteries at much lower temperatures and with less toxic output than standard battery manufacturing processes.

To succeed in this project, Belcher first had to solve two challenges. First, she had to figure out how to organize metal battery materials to pack as densely as possible. But simply gathering together battery materials wasn't sufficient, because electrons and metal ions need to find efficient ways to travel through and around the battery materials. So she also had to engineer conductive channels for electrons to travel from an external source through the battery electrodes. She needed a nanometer-sized particle that could both bind metal ions and provide conduits for electrons, and M13 offered many of the features she needed.

The M13 virus looks like a tube—an exceedingly tiny, exceedingly thin tube, decorated at each end with thread-like tufts. A single M13 virus measures a bit shorter than 1,000 nanometers (nm) in length and a little narrower than 10 nm in width (1,000 nm is about one-tenth the width of a human hair). The M13 proportions resemble a very long version of a Twizzler licorice candy stick, six Twizzler sticks long and only one Twizzler in width. The M13 tube twists, a bit like the surface of a Twizzler, and is constructed from about 2,700 copies of a single protein called p8. The p8 proteins are arranged in a very regular compact array. Belcher recognized the awesome packing potential of M13: if each of those 2,700 p8 coat proteins could be engineered as a binding site on which to anchor a critical battery structure, her M13-based electrodes could charge and discharge with exceptional speed.

Belcher has used the full gene engineering toolkit developed

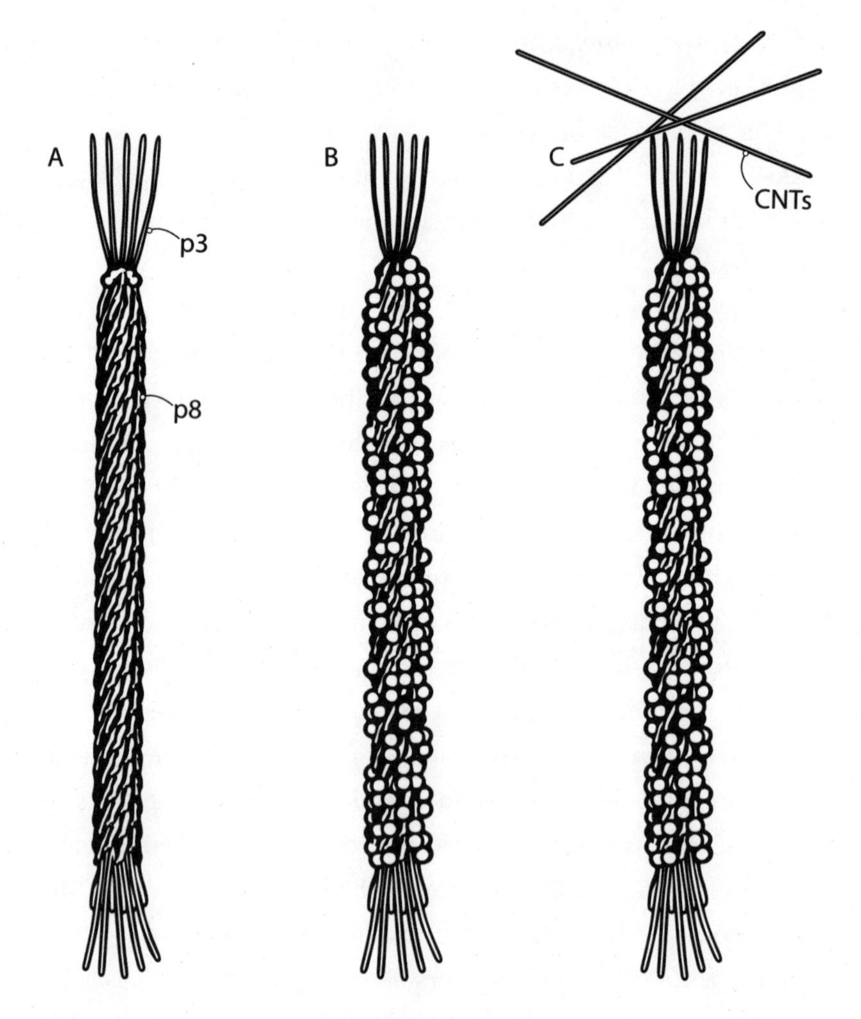

The M13 virus can be modified to bind to battery materials. (A) The M13 virus has a long, cylindrical shape with a core composed of 2,700 copies of the p8 protein. One end carries strands of p3 proteins, which normally mediate attachment to a host cell. (B) Variants of M13 with modifications of their p8 proteins allow them to bind to battery materials, such as cobalt oxide (small spheres). The cobalt oxide particles decorate the cylindrical core. (C) M13 can be further modified so that the p3 proteins bind single-walled carbon nanotubes (CNTs).

by biologists to modify the M13 virus to make better batteries. Initially, to identify variants of M13 that could most densely pack battery materials together, she adapted a technique called phage display, which was originally developed to study the molecular components of the immune system. Belcher started by mutating M13 to make a billion distinct M13 variants, each having a slightly different genetic sequence. Hypothesizing that among those billion she would find some with the appropriate properties, she began testing them for the ability to bind to interesting materials to which viruses do not typically bind, such as gold or carbon nanotubes. Through several rounds of producing M13 variants and selecting those with promising interactions, she identified several that bind tightly to these materials.

Belcher next went a step further to make M13 even more adaptable. Reasoning that if she could redesign the 2,700 copies of the native p8 coat protein to bind to other general classes of molecules, she would have a multifunctional tool, she and her colleagues added a gene sequence that would give the p8 protein a string of negative charges. This gave every p8 protein a sticky end that can hold onto positively charged particles—battery materials like cobalt oxide, for example. Just imagine the power of this approach: the new Belcher-modified M13 virus had 2,700 sticky ends, equipping it with sites for binding positively charged metal particles.

Belcher didn't stop there. Getting 2,700 molecules of M13 to provide attachment sites for battery components was a huge advance, but she also needed to make sure that the electrons and ions that must move through the electrodes could flow speedily. To tackle this problem, she switched her attention to another M13 protein, the p3 protein that makes up the thread-like strands at one end of the M13 central tube. In the natural world, the p3 proteins of the M13 virus bind to the surface of its host, the *E. coli* bacterium—hence, M13's name, *bacteriophage*. Belcher reasoned that if p3 can bind to bacteria, perhaps she could modify it to bind to materials that would serve as conduits for the electrons and charged metal ions that must race through batteries. She and her colleagues again used phage display to identify p3 variants that could bind to one of the best-known conductors of ions, single-walled carbon nanotubes.

All of this work eventually made it possible for Belcher to create a library of very specifically modified variants of the M13 virus. Each variant bound one or two materials that are useful in building batteries. Some modifications mediated specific interactions with materials like gold; others mediated nonspecific interactions with ionic materials so that they could interact with charged particles like cobalt oxide or iron phosphate. Some bound carbon nanotubes to accelerate electron transport; others were engineered into a two-gene system to create super variants of M13. With these new tools in hand, Belcher's lab started to produce virus-based battery electrodes.

o

I wanted to see these virus factories with my own eyes, so I visited Belcher in her lab, where she assigned Alan Ransil, one of her battery-enthusiast graduate students, as my guide. "Enthusiast" doesn't begin to describe Ransil's excitement about the future of energy storage or how eagerly he shared his understanding and ambitions with me.

When Ransil opened the door to the Belcher lab, every bench was packed with advanced technology machines and the tools and materials to use them. A steady stream of graduate students and postdocs walked purposely from machine to machine, in and out of the different bays and rooms. Angie Belcher's work attracts young researchers from all over the world and from a dozen different disciplines; her lab hosts close to twenty researchers at any time, with each staying anywhere from six months to several years. Ransil did his undergraduate research at Stanford University where he focused on developing new materials for solar cells, but now he's the resident expert in designing batteries into new shapes like watchbands or car dashboards. Geran, from New Zealand, has a background in materials engineering and now works on designing sulfur-based, high-capacity battery electrodes. And Nimrod, from Israel, with a background in biology, now works on 3D printing of bacteriophages for batteries. It's a United Nations of budding energy professionals, with degrees in applied physics and chemical engineering, biology, and materials science, from Turkey, India, Japan, America, China, Canada, the United Kingdom, Germany, Korea—and beyond. While they work with an incredible sense of purpose on their particular projects, there's no telling what possibilities might emerge from their hallway conversations. What if you could engineer viruses to turn natural gas into gasoline? Is it possible to invent a new way to visualize small groups of tumor cells to make cancer surgery more effective?

Most of these "crazy" ideas rapidly get discarded, but several have progressed beyond the lab. One Belcher lab spin-out company, Siluria, converts natural gas into gasoline and other liquid fuels, promising a cheaper way to transport and store methane and gas. Another project from the lab recently moved into a clinical trial to test whether a novel imaging technology that Belcher and her colleagues have designed can more effectively guide surgery for ovarian cancer and improve patient survival.

As I toured the lab, Ransil led me from room to room,

demonstrating the steps to construct a virus-made battery. Walking into a room full of freezers that store Belcher's virus libraries, he opened one of them, revealing dozens of five-inch-square boxes. He took one box from the freezer, removed one of 144 carefully arranged vials from the box, returned the box to its shelf, and closed the freezer door. He worked quickly to minimize temperature changes in the freezer, which is maintained at –80 degrees Celsius to prevent degradation of the samples in the virus libraries.

Earlier, Ransil had prepared a colony of host bacteria for viral infection. I watched as he carefully thawed the frozen virus sample and then added it to the bacteria, allowing the viruses to infect the bacteria. He would expand the infected culture of bacteria over the next twelve hours by, first, transferring it into a growth medium in a small flask swirling on a shaker table at 37 degrees Celsius (human body temperature). He'd then move it to a larger vessel and then to a much larger vessel. After that, he would purify the now 10^{16} amplified viruses from their host bacteria. He showed me the various steps, moving material from lab benches at room temperature to cold rooms to stirred solutions on heated plates, each step carefully calculated to purify the components and then adding them together at exactly the right moment in exactly the optimal concentration. As we walked from one lab station to the next, we paused frequently at a diagram on the hallway wall to track the steps in virus-based battery making. The whole process seemed a bit like following a cookbook, except that Ransil and his colleagues have written all the recipes themselves, and they constantly work to improve them.

At the end of the mixing, growing, purifying, melting, weighing, and drying stages, it was time to assemble the battery. We entered a lab room, where we were greeted by a phalanx of black rubber arms and hands that extended, fingers outstretched, from a long, glass-enclosed chamber that housed a lab bench. Ransil explained that the rubber arms were actually gloves, inflated by a constant pressure of argon gas that fills the chamber. Argon is nonreactive and relatively cheap, and it keeps the environment of the chamber essentially free of the oxygen and humidity of ambient air, which would destroy the battery components before they are assembled into batteries.

Ransil pushed his hands into a pair of the gloves, inverting them into the chamber so that he could work inside the argon-filled box. Using a pair of tweezers, he assembled a stack of battery components into a flat, circular battery case. He placed one side of the case onto an ultraclean sheet of lab paper and started to build the battery. The first layer was a disc of lithium foil that served as the anode. He then added a few drops of electrolyte solution, then a plastic separator and another few drops of electrolyte—and then another disc that looked like a piece of foil but was, in fact, the virus-based battery cathode.

He added a few more drops of electrolyte, closed the battery casing, crimped it shut, and pronounced it a battery.

Ransil's battery looked just like the coin-shaped batteries I've handled when changing the battery in my watch. And on the outside, they are the same. The Belcher lab packs novel, biology-based

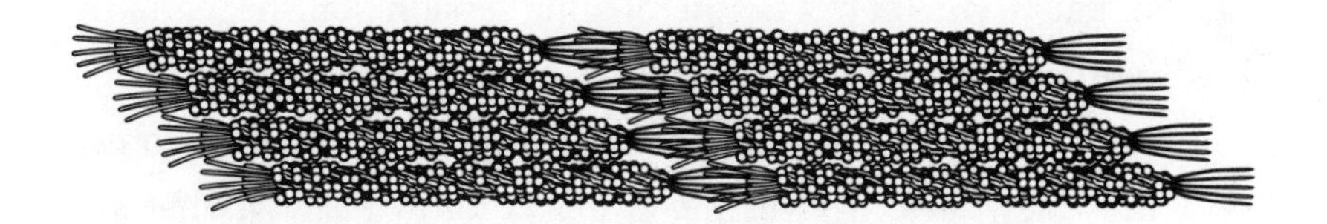

The structure of M13 facilitates self-assembly into sheets of electrode material.

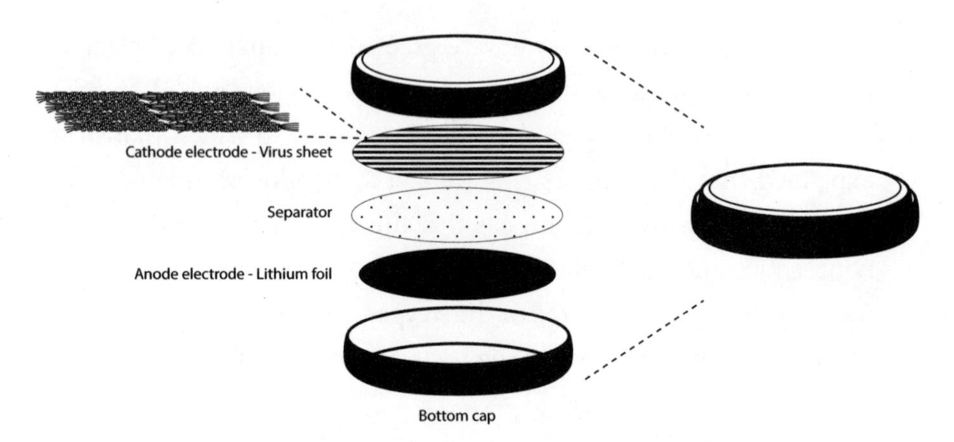

Assembly of a virus "coin cell" battery. A layer of lithium foil serves as the anode, and a layer of modified M13 viruses serves as the cathode. All the layers of the components are sealed together in a coin cell battery case.

battery components into standard battery casings, which then can power conventional electronics.

I have always loved laboratories. I love the sights, smells, and machines. But most of all, I love the intensity of the work and the shared spirit of collaboration that makes impossible things possible. The Belcher lab reminds me of my neurobiology lab in many ways, but my mind spins when I try to comprehend the engineering that Angie Belcher has brought together with biology in such unusual and unpredictable ways.

One recent afternoon, as we talked about the future of energy, Belcher had to run off to a brainstorming session that she regularly holds with members of her group. "It's my absolute favorite thing," she said. "When we put our heads together and someone comes up with a new idea, it gives me chills." I know exactly what she means: the magic of collaborative thinking. Belcher's unique kind of cross-disciplinary thinking is the stuff of genius, and Belcher has a gift for fostering it, which is why she received a MacArthur Fellowship, the "genius award," in 2004.

o

Step by step, Belcher has used her new virus-enabled tools and techniques to assemble all the components necessary for a battery. In 2006, she reported the successful construction of a virus-enabled anode, and in 2009 she did the same for a cathode. The idea that viruses can be used to improve both of the energy-storing components of a battery attracted widespread attention. When President Barack Obama visited MIT in the fall of 2009 to highlight his national commitment to inventing a sustainable energy future, we showed him several promising new energy technologies, including the virus-enabled batteries. After Belcher explained to the president her goal of discovering new materials for her pioneering biological-fabrication processes, she handed him a pocket-sized periodic table, explaining that he could use it "if you're ever in a bind and need to calculate molecular weight." Without missing a beat, he responded, "Thank you, I'll look at it periodically."

Battery production today is an energy-intensive process that produces considerable toxic waste. Like the abalone shell, however, Belcher's virus-produced batteries assemble benignly. This represents an important contribution to solving our energy-storage challenge, and Belcher is rightly proud of the work she and her colleagues have done. "These biological batteries," she told me, "are all made at room temperature, use no organic solvents, and add no toxic materials to their environment." Compared to standard battery-manufacturing processes, which can require temperatures of close to 1,000 degrees Celsius and generate about 150–200 kilograms of carbon dioxide–equivalent waste per kilowatt-hour of battery in the cell manufacturing phase, Belcher's approach offers a massive advance on our energy-storage challenge.

But Belcher isn't stopping there. Her next question is whether her state-of-the-art, virus-based batteries can do more than transport and store energy. Rather than passively add weight to a car, could they take the form of a dashboard, a seat cover, or a door panel? That could prove to be the "killer app" that moves virus-based batteries from her MIT laboratory into the marketplace, following the path of her other lab-based start-up companies.

Belcher has confidence in an energy future that's fundamentally different from the present, and she shares this confidence with many pioneers on the frontier of energy innovation. Our energy economy, she recognizes, will not always depend on oil. Even Sheikh Yamani, Saudi Arabia's oil minister from 1962 to 1986, a period when world crude oil production more than doubled, understood this truth. "The Stone Age," he once said, "came to an end not because we had a lack of stones, and the Oil Age will come to an end not because we have a lack of oil."

We're still in the Oil Age, of course. But by harnessing the intelligence of biology, Angie Belcher and her colleagues believe that they can help bring it to an end.

3

WATER, WATER EVERYWHERE

In the late 1980s, Peter Agre stumbled on a discovery that would change the way we think about water. As a newly appointed physician-scientist in the Hematology Division at the Johns Hopkins University Medical Center in Baltimore, Agre wanted to get his hands on the protein that caused Rh disease, a disorder that has the terrifying potential to damage a developing fetus. Red blood cells carry the Rh protein on their surface, and when the Rh protein on a mother's red blood cells doesn't match that of her developing baby, the mother's immune system attacks the Rh protein on her baby's red blood cells. That immune attack can kill the developing baby's red blood cells, depriving the baby of oxygen and causing a host of problems—sometimes even death. Although enormous progress had been made in protecting babies from Rh disease, no one had yet identified the Rh protein or determined its normal function.

Agre resolved to figure these things out. He followed a classical strategy, purifying enough Rh protein from red blood cell membranes to identify it, once and for all. Starting with a large volume of red blood cells, he separated the cell membranes from the rest of the preparation. Next, he designed a careful set of steps to isolate the Rh protein from other proteins present in red blood cell membranes. But as he moved to the final step, much to his dismay and consternation he discovered an interloper—a contaminant that had traveled, undetected, alongside the Rh protein through his elaborate purification strategy. No matter how carefully he worked, every time he ran his experiment, the contaminant made it through.

It was maddening. All lab scientists know the feeling. You take every precaution, you make every check and cross-check, and then your exquisitely pure sample ends up, well, not so pure. First you don't believe the results. Then you suspect a problem with the lab protocol. Then you feel a stomach-churning sense of defeat. Eventually, though, you start working your way down a long list of possible explanations. That's what Agre and his colleagues did. Initially, he hoped for the best-case resolution, that the contaminating protein was a fragment of the Rh protein. But his further analysis showed, disappointingly, that the contaminant was not a fragment of the Rh protein but was some other previously unidentified protein. Agre had no idea what it was or what it did. And he certainly had no idea that in isolating the interloper he had made a discovery that ultimately would win him the 2003 Nobel Prize in Chemistry and that would open up revolutionary possibilities for purifying the world's freshwater supply.

o

We cannot live without water. It makes up more than 50 percent of our bodies, and we rely on a ready supply of it for drinking, agriculture, transportation, manufacturing, and more. Water is everywhere: some 300 million trillion (300×10^{18}) gallons cover about 70 percent of Earth's surface. But almost all of that—more than 95 percent—is salty ocean water, which we can neither drink nor use to water crops or to meet most of our water needs.

We need freshwater to survive, but freshwater makes up less than 5 percent of the planet's total water volume. And most of that freshwater sits in ice sheets, the soil, and the atmosphere. Only about 1 percent of Earth's freshwater is accessible for our use, and it's already not enough to support life as we know it. More than 1 billion people lack access to potable water today, and drought haunts both the developing and the developed worlds. We need more freshwater; the obvious way to get it is by purifying the salt water and contaminated water that exist in great abundance all around us.

Water purification has been critical for human survival for a very long time. Paintings from ancient Egypt as early as 1500 BC illustrate water purification by filtration, and Aristotle described purification by distillation. Although we've gotten much better at water purification since then, we still primarily rely on those same two basic technologies. And even after four thousand years of refinement, water purification by distillation and filtration is too slow, too expensive, and too energy inefficient to meet our growing demands. We need dramatically new approaches to purifying water. And Peter Agre's 1992 discovery made possible a tantalizing new possibility, even though he didn't realize it at the time. The answer to our water challenge, it turns out, might lie inside our own bodies, in Agre's mystery protein.

o

In 1988 Agre published a paper reporting the identification of his new red blood cell protein. In that paper he confessed that its role was "uncertain," a humbling statement for any scientist to make.

Agre puzzled over what this mystery protein might do, but he didn't make much progress in solving the puzzle until a family camping trip in 1991.

Agre and his family love the outdoors and they often spent their vacations camping in national parks. That year, when he and his wife asked their children what national park they should visit, their children's choice was instantaneous and unanimous: Disneyworld! So Agre and his wife made Florida that year's destination. Rather than follow their children's wishes exactly, instead, like all good parents, they packed up their car and headed for Everglades National Park. They did, however, concede to their children's choice of destination by breaking up their long return trip with a stop at Disneyworld, camping in Jellystone Park. After that, while on their way back to Baltimore, they decided to stop at the University of North Carolina (UNC), Chapel Hill, to visit Agre's old friend and mentor, Dr. John Parker. It was a fortuitous decision.

Parker, a clinical hematologist and oncologist, had supervised Agre during his clinical training. As often happens, Parker had continued to serve informally as Agre's trusted mentor. It was Parker who had encouraged Agre to study red blood cell membranes. During their post-Disneyworld visit, Agre couldn't get his mind off the lab puzzle and shared his confounding results with Parker. He described how the mystery protein had invisibly followed the Rh protein through the elaborate and time-consuming purification strategy and how he had found it abundantly expressed in the kidney, which did not have the Rh protein. Try as he might, Agre could not figure out what the mysterious protein could be. As he told Parker the saga, it didn't take Parker long to figure it out. Recognizing that kidney cells and red blood cells both transfer a lot of water across their membranes, Parker deduced that Agre may have discovered the long-sought but stubbornly elusive water channel, and that Agre's mystery protein might provide the answer to a question that had long eluded scientists: How does water cross cell membranes?

Scientists had long recognized the importance of this question. Each of the roughly 35 trillion cells that make up our bodies carefully monitors and regulates how much water passes through its membrane. Some researchers theorized that a unique channel protein for water had to exist, but despite an enormous effort, no water-transferring protein had been found. Hematologists had a particular interest in how cells achieve the right balance of water between the inside and the outside of a cell, because red blood cells need to maintain just the right amount of internal water to do their job of ferrying oxygen from the lungs to all the body's tissues and then returning carbon dioxide back to the lungs. Only when plumped up with water can red blood cells carry their vital payload. So you might expect to find water channel proteins in abundance in red blood cells, if such proteins existed.

Recognizing the importance of water regulation for cell viability, many scientists had attempted to identify a water channel. Water flows passively into and out of cells by a process called osmosis, with its direction of flow determined by the concentration of what's dissolved in the water on either side of the cell's membrane. Osmosis balances the concentration of solutions on two sides of a membrane or filter. Simply put, if a water-permeable filter separates pure water on one side from salty water on the other side, the pure water will flow through the filter to dilute the salty water until the concentration of salt is balanced on both sides.

But how did water move across a cell's membrane? At the time Agre and Parker met up in 1991, most researchers had decided that water didn't need a special pore or channel to ferry it into and out of cells. The accepted model was that water naturally diffused across cell membranes, just as it diffuses across other filters.

Despite this accepted model of water transfer, Parker's noncanonical idea—that is, that the mystery protein could be a water channel—intrigued Agre. But he wondered if it was worth pursuing. Could he really disprove a well-accepted scientific model by studying a protein that most people didn't believe existed? The effort would probably be a wild-goose chase, he knew, as it had been for the many other researchers who had fruitlessly pursued the water channel. And that pursuit would undoubtedly divert Agre from his research on the Rh protein. The wise thing to do was just to drop the idea. But he couldn't. Instead, he decided to go on that wild-goose chase.

Pursuing the mystery protein required Agre to change the direction of his work in the lab. To prove that the mystery protein was, indeed, a water channel, he decided to test the protein's function in a different kind of cell, one that did not normally move water across its membrane. Agre and his colleagues identified specific DNA strands that coded for the mystery protein and then copied the protein's DNA into RNA. The RNA, injected into another cell, would direct that cell to make the mystery protein. Agre's strategy was to get a test cell to make the mystery protein, to determine whether, as Parker proposed, the protein would create water channels and transport water across the test cells' membranes.

Agre used frog eggs to test Parker's idea. Why frog eggs? Because he knew that frog eggs remain plump and full of everything they need to nurture a developing frog when submerged in fresh pond water for days on end. Even with their very high concentration of salts and proteins inside, frog eggs seemed impermeable to water, which suggested that their soft enclosing membrane had no mechanism for transporting water in and out.

Agre designed a relatively straightforward test. First, he injected the mystery protein's RNA into one batch of frog eggs and, as a control for manipulating the eggs, he injected water into another batch. The eggs injected with the RNA, he reasoned, should direct those eggs to produce the mystery protein. After a few days sitting in a salty saline solution, both sets of eggs looked the same. But then came the test: he placed both sets of eggs in pure water. The control eggs behaved like frog eggs—nothing happened. But the eggs that made the mystery protein, he told me with delight, "exploded like popcorn."

What was the difference? Agre could draw only one conclusion. The mystery protein's RNA had made water channel proteins, which had inserted into the membranes of the eggs. When the saltiness inside and outside of the eggs was balanced, both batches of eggs looked the same. But when the eggs were placed in pure water, the water channels in the RNA-injected frog eggs allowed water to flow into the cell, filling them to the point of bursting.

Proof! Through both serendipity and brilliant detective work, Agre had found the elusive water channel. He named it aquaporin. Soon it became clear that he had discovered only the first of what we now know is a whole family of aquaporins, found in virtually every organism on Earth: in animals and plants, in bacteria and fungi.

o

Agre's brilliant biology puts the water channel in the hands of not only scientists but also engineers and entrepreneurs, some of whom now hope to deploy it for large-scale water purification. To understand how a water channel works for a cell and how it might be repurposed to purify water for our use, we need to think about what proteins are and how they work.

I like to picture proteins as mini-machines, each designed to do a highly specific job for a cell or a tissue. Mechanistic analogies can help us understand these jobs. Aquaporin functions a bit like a parking-lot gate that permits only cars identified with a specific transponder to enter. The structure of the aquaporin channel, or gate, recognizes the atomic signature of water and allows only a molecule with that particular signature to pass into and out of a cell. It blocks the passage of salt, acid, and all other molecules.

However, unlike a gate, the components of protein machines aren't forged from metals or molded out of plastic. Proteins are strings of beads, strung together in a highly precise order.

The beads in these protein strings are molecules called amino acids, which come in twenty-one different varieties. These strings of amino acids wind themselves into highly ordered structures that form the parts of a protein machine. Two important features give each protein its specific structure and function. First, the amino acid beads in each protein string line up in a unique order. There may be only twenty-one amino acids, but if you consider that a standard protein might contain more than a hundred amino acids, their possible combinations are legion. Second, because some amino acids attract each other while others repel, these forces of attraction and repulsion cause each amino acid string to wind itself into a specific shape; it is this shape that allows a protein to carry out its function.

Proteins perform a variety of functions. One protein family provides conduits for materials to cross cell membranes. These conduits, or channels, are highly selective, ferrying only one or a small set of molecules into or out of a cell. Some channels move their cargo through one-way gates that allow their specific molecular cargo to pass in only one direction into or out of a cell. Others are two-way gates, carrying their specific cargo in both

Proteins are composed of strings of amino acids, resembling beads on a string. The order of the amino acids is determined by the order of nucleic acid bases in DNA (or RNA). The chemical properties of each of the twenty-one standard amino acids determine the structure of the protein through attraction and repulsion with the other amino acids in the protein string. Those forces can make a segment of a protein wind up into a helical shape (middle of protein string), fold into a sheet (bottom of protein string), or take on other conformations.

directions. Some channels ferry sodium, some ferry chloride, and some—as Agre demonstrated with his discovery of aquaporin—ferry water.

Once he showed that his mystery protein functioned as a water channel, Agre embarked on an exploration of the new world of aquaporins. After he determined the full sequence of amino acids in the aquaporin protein string, he then determined that the string winds up and loops around into a configuration that resembles an hourglass with a very narrow neck. The hourglass spans the thickness of the cell membrane, and its central opening serves as a highly selective channel to ferry water bidirectionally between the inside and the outside of the cell.

The precise winding of an amino acid string positions particular amino acids in particular positions. Different amino acids have different properties: some are positively charged and some are negatively charged; some repel water and some attract water; some repel fatty substances, such as the lipids that form biological membranes, and some attract them. Agre and his colleagues showed that the amino acids in the aquaporin amino acid string are positioned so that amino acids that attract lipids form the outside surface of the hourglass (and interact with the lipid cell membrane), and amino acids that attract water line the internal surface of the hourglass.

But what was it about aquaporin that allowed water to pass into and out of a cell while preventing everything else from passing through? Agre mapped the amino acids that line the channel opening and discovered that the walls that line the channel's pore have alternating negative and positive charges that ferry water molecules—and only water molecules—through the channel.

The secret of aquaporin's specificity for water rests on the atomic structure of water. Water molecules have an asymmetric structure with an asymmetric distribution of charges. An individual water molecule (H_2O) consists of two atoms of hydrogen and one of oxygen. Its single oxygen atom makes one side negatively charged, while the two hydrogen atoms make the other side positively charged. In high school science classes, we learn that a water molecule's asymmetry accounts for water forming a crystal in its solid form as ice through the complementary attraction of

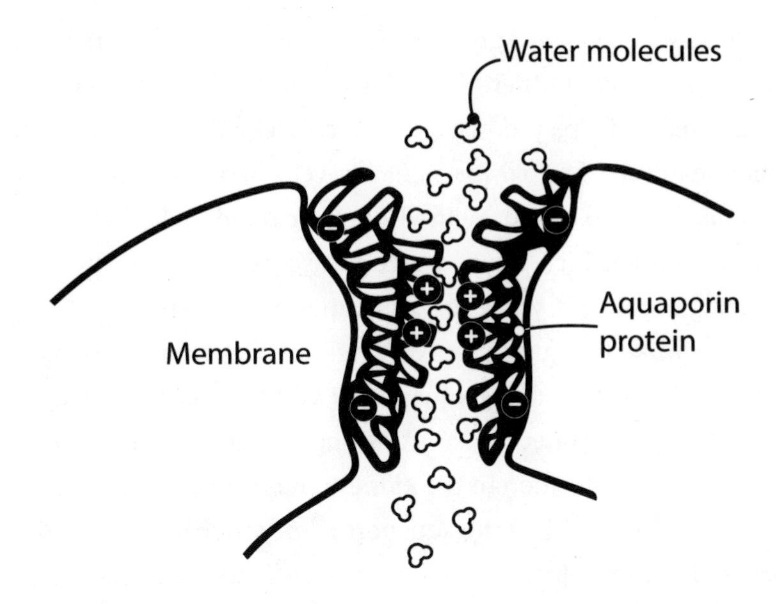

The aquaporin protein forms hourglass-shaped channels in cell membranes. These channels allow water to cross the fatty cell membrane layer. Viewed in cross section, the aquaporin amino acids facing the central pore of the water channel attract water, whereas the membrane-facing side of the aquaporin protein is composed of lipid-attracting (membrane-attracting) amino acids. The distribution of negative (–) and positive (+) charges conducts water molecules through aquaporin's channel.

its positive and negative charges. The asymmetry in charge of the water molecule comes into play here, too: the alternating negative-positive-negative charges along the aquaporin pore's lining escort water molecules, one by one, through its channel at an astonishing rate of up to 3 billion per second.

After Agre identified the first aquaporin, he and other researchers soon found other members of the now very large family of aquaporins. Almost every organism has them—from simple bacteria to complex plants and animals. Some conduct only water, like the first aquaporin Agre discovered, and some members of the family also conduct other molecules, such as glycerol and urea.

Agre tends to play down his discoveries. "There wasn't any genius involved," he told me, "but I'm very happy to have solved it, using the time-tested method of sheer blind luck." That's too modest. Perhaps luck played a role but so did determination, an inquiring mind, and, yes, a very large measure of genius.

In the wake of Agre's discovery, researchers from all sorts of disciplines began studying aquaporins and the roles they play in processes such as water transport through plant roots and rootlets as well as water filtration in the kidney. These advances have been remarkable. But in 2000 the aquaporin story took an entirely different turn when Morten Østergaard Jensen, then doing his doctoral studies in biophysics at the University of Illinois, read Agre's report on the atomic structure of aquaporin. As he read, Jensen felt a light come on in his mind: What if it were possible to use aquaporin to purify water not just for cells inside living things? What if it were possible to make an aquaporin-based water filter to purify water for all of us to help meet civilization's growing demand for clean water?

o

To investigate the potential of his idea, in 2005 Østergaard Jensen teamed up with his friend Peter Holme Jensen, a serial entrepreneur with a scientific background in structural biology. After the two understood more fully how aquaporins work, they had greater confidence that it might indeed be possible to make an aquaporin-based water filter by incorporating the protein into a membrane sheet to create a kind of sieve. That sieve, they imagined, would allow water—and only water—to pass through. The more they thought about the idea of a biologically based filter, the more potential they felt it had. Aquaporin's specificity for transferring water molecules could make it more efficient than currently manufactured filters. As Holme Jensen would later put it to me, "Why not use nature's genius rather than trying to invent something better?"

In 2005 the two men launched Aquaporin A/S, a water-purification company based in Denmark, with the goal of harnessing the selectivity of aquaporins to engineer a new kind of water purification technology. They knew well that meeting the looming water needs of a planet that in 2050 will be home to more than 9.5 billion people will require new technologies. Together, the two decided to see what they could do to develop a new water purification strategy. When I asked Holme Jensen how he made the startling intellectual leap from scientific discovery to a possibly world-saving application, he replied, "It was obvious." Obvious to him, perhaps—but he now had the job of convincing and recruiting others to help him pursue his visionary idea.

Holme Jensen made a hiring breakthrough in 2006, when he enlisted his colleague Claus Hélix-Nielsen as the chief technology officer of Aquaporin A/S. At the time, Hélix-Nielsen held—and has continued to hold—faculty positions in the Department of Environmental Engineering at the Technical University of Denmark and the Department of Chemical Engineering at the University of Maribor Slovenia. It was a catalytic match. Together, they embarked on an adventure to figure out if they could scale up an aquaporin-based water purification membrane from meeting the water needs of a cell to meeting the water needs of a city.

To learn more about the work Aquaporin A/S is doing, I flew to Denmark to visit Hélix-Nielsen. He's a man of wide-ranging interests; as we discussed aquaporins, he veered into all sorts of fascinating territory, treating me to reflections on such matters as the physical limits of optics in animals and in machines as well as the written correspondence between Mozart and Haydn. As a boy, he thought he'd become an archeologist or perhaps an architect. But then he became preoccupied with understanding how the human brain processes and stores information. So he shifted his studies to biophysics. At first, he thought he might study the human visual system. But in trying to figure out how the brain's nerve cells and their complex circuitry produce vision as we perceive it, he became intrigued by the more general problem of how the individual activities of the components of a system assemble to carry out the complex activities of the system as a whole. He soon realized that human vision was far too complex a subject if he wanted to make progress on that general problem, so he focused on something straightforward: how materials move across cell membranes through specialized protein channels and how these proteins interact with the membranes. Eventually, in 2005, he found himself drawn to aquaporin, captivated by the ease with which aquaporin separates water from every contaminant.

Of course, for an aquaporin-based water purification system to work practically, it would have to hold at least the promise of matching the cost and efficiency of current systems. Could such a device be designed and developed to solve the need for pure water on a global scale? As soon as I arrived at the Aquaporin building, Hélix-Nielsen walked me to a display area where an intriguing array of plastic cylinders of various sizes showcased the company's products. The cylinders, labeled with the logo "Aquaporin Inside," were water filtration devices, ranging from a few inches in length to over three feet. Most were still in development, but Hélix-Nielsen proudly handed me one, about a foot in length and four inches in diameter, that the company was already testing in private homes in China.

o

Hélix-Nielsen, Holme Jensen, and their colleagues at Aquaporin A/S faced a set of daunting challenges as they contemplated scaling up water filtration from the cellular to the city level. A red blood cell makes enough aquaporin only to filter water for its own needs—which, given the size of a single cell, aren't great. A red blood cell measures less than 10 micrometers (μm) in diameter, which means that 150 red blood cells lined up end to end would be only about as thick as a dime. Hélix-Nielsen and his colleagues recognized that if they wanted to create a commercial filter of any kind, even if just a home filter to start, they would have to find a way to produce a whole lot more aquaporin than cells do. The aquaporin they produced would have to be stable because commercial filters don't have the machinery that cells do to replace aquaporin that wears out. And because they were trying to build a filter that could handle much greater water flow than occurs in cells, the aquaporin they made would have to be positioned in a membrane that was much stronger than a cell membrane.

To tackle the first problem—namely, producing proteins in a quantity suitable for commercial use—Hélix-Nielsen looked to the biopharmaceutical industry for ideas. He couldn't have done this in the twentieth century. The blockbuster drugs of the twentieth century, among them aspirin, acetaminophen (Tylenol), atorvastatin (Lipitor), and omeprazole (Prilosec), came out of chemistry labs, which devised ingenious methods to identify and then chemically synthesize a large variety of chemicals that can intervene in or supplement cellular processes with very high specificity. But many of our newest drugs are biological products, not synthetically manufactured chemicals—for example, proteins harvested from growing cells. By harnessing the intelligence of biology, a new breed of drug discoverers has manipulated biological mechanisms to persuade living cells to produce protein-based drugs like adalimumab (Humira) and etanercept (Enbrel) for autoimmune diseases and a host of new anticancer drugs, including trastuzumab (Herceptin), rituximab (Rituxan), and bevacizumab (Avastin).

To produce enough of these drugs for human consumption, the biopharmaceutical industry has figured out how to grow microorganisms in huge vats and then how to purify specific protein drugs from them. Companies like Genentech, Genzyme, Amgen, and Biogen have developed new methods to reliably ramp up the production of this new class of drugs to industrial scale. Hélix-Nielsen reasoned that it would make sense to use the same techniques to mass-produce aquaporin. In 2006 he and his colleagues teamed up with molecular biology experts and began working with various microorganisms to find a cellular "factory." The factory they chose was the bacterium *E.coli*, and they did so for two reasons. First, it's an easy organism to grow, and they knew that the biopharmaceutical industry had already figured out how to deploy it as a bio-factory to make protein-based drugs such as insulin and growth hormone. Second, *E. coli* naturally makes its own aquaporin, so it should be possible to persuade it to make more without killing the cells. All in all, *E. coli* seemed to have the potential to become a robust producer of aquaporin for commercial applications.

But Hélix-Nielsen and his team confronted additional challenges. A critical difference exists between aquaporin and current biopharmaceuticals. The cells producing most biopharmaceutical drugs today do their work by secreting the protein drugs into the fluid around them. These secreted proteins can then be collected with the fluid that bathes the cells and purified from that bathing fluid. But, as Agre had shown, aquaporin sits in cell membranes. Cells that produce aquaporin don't secrete it into the surrounding fluid; they insert it into their membranes. To harvest aquaporin from cells, the Aquaporin A/S team would have to first purify membranes and then develop membrane-busting techniques to extract the aquaporin, two harsh processes that many proteins can't survive.

As we've seen, a protein's function depends on how a string of amino acids winds up into and then maintains its specific three-dimensional shape. This winding-up process relies on forces of attraction and repulsion between amino acids; if those forces are disrupted by any one of a number of factors, a protein can unwind, creating structural changes that can interfere with the protein's function. A protein's function also can be disrupted if the protein string is broken, at which point a protein can't do its job. In a living cell, these generally are not terrible problems because the cell machinery can repair or replace defective proteins. But for a protein to be useful in, say, a commercial water filter, it needs to maintain its structure for a long time without repair or replacement. Fortunately, aquaporin is unusually stable.

Perhaps it has to be. Unlike most of the other cells in our bodies, red blood cells don't carry the machinery to repair and replace their components, and they also lack the machinery to divide. As a result, red blood cell proteins need to be very stable to withstand the arduous journey they make through the body's vascular system. They also must be stable enough to live for about four months, which is a long time compared to skin cells that live less than a month or to cells lining the digestive system that live less than a week. For the Aquaporin A/S team, this was good news.

Because of how stable aquaporin needs to be to serve red blood cells, it can survive the extreme temperatures and chemical conditions required to purify it from cell membranes. Even after being subjected to a tough purification process, aquaporin does not lose its water-filtering powers. For this, Hélix-Nielsen told me when describing their adventure in building an aquaporin water filter, "We are blessed."

o

So far, so good. Modifying techniques from the biopharmaceutical industry, the Aquaporin A/S team had designed a process to manufacture lots of aquaporin. But the next challenge loomed: that is, how to re-embed the isolated aquaporin proteins in membranes.

Solving that problem turned out to be relatively straightforward. Aquaporin's structure makes it seek out a lipid environment—that is, an environment made up of molecules that are more like oil or butter than like water. A cell membrane is a lipid environment, one that keeps the water environments inside and outside the cell separate. The internal and external environments contain different molecules in solution: inorganic molecules like sodium chloride (standard table salt) and other salt-like molecules, and organic molecules like proteins that dissolve in water. Each environment, inside and outside the cell, needs to maintain the correct composition of molecules in solution, and the cell membrane makes that possible. But as required, aquaporins and other channel proteins will transfer water (in the case of aquaporin) or sodium or other molecules (in the case of other channel proteins) from one side of the membrane to the other.

Recall that the barrel walls of the aquaporin hourglass, like all parts of a protein, are constructed from amino acids. Some amino acids dissolve in water (they are hydrophilic, or water loving) and some amino acids dissolve in fat and avoid water (they are hydrophobic). The outside surface of the aquaporin barrel is composed of hydrophobic amino acids that seek lipid environments. Given all this, once Hélix-Nielsen and his team had prepared purified aquaporin, all they had to do was to mix the aquaporin with lipid-like membrane spheres they had constructed from special polymers, and the aquaporins naturally inserted themselves into the spheres.

The next step was to construct a filter from those very small, aquaporin-containing, polymer-based spheres. The obvious strategy would be to open up the spheres and allow them to connect to one another in a flattened sheet. But membranes naturally want to be spheres. Flattening them, and keeping them flat, is a challenging and costly process. In 2009 Hélix-Nielsen and his team began a collaboration with scientists from the Singapore Membrane Technology Center to determine whether they could dispense with the flattening process altogether and make a filter that consisted, improbably, of membrane spheres—or, as he calls it, a vesicle sheet. Instead of a flat membrane with aquaporin pores, they imagined a thin layer with aquaporin spheres embedded in it.

This created a potential problem: to pass through a sheet of intact vesicles required that the water molecules would have to make two aquaporin passages rather than just one. Water would have to enter the spheres through an aquaporin channel on one side of the sphere and exit through another aquaporin channel on the other side.

The next big question that the Aquaporin A/S team needed to answer was whether having these two passages, instead of one, would reduce the efficiency of the filtration process. The answer, it turned out, was no. In 2012, Hélix-Nielsen and his collaborators at the Singapore Membrane Technology Center managed to develop

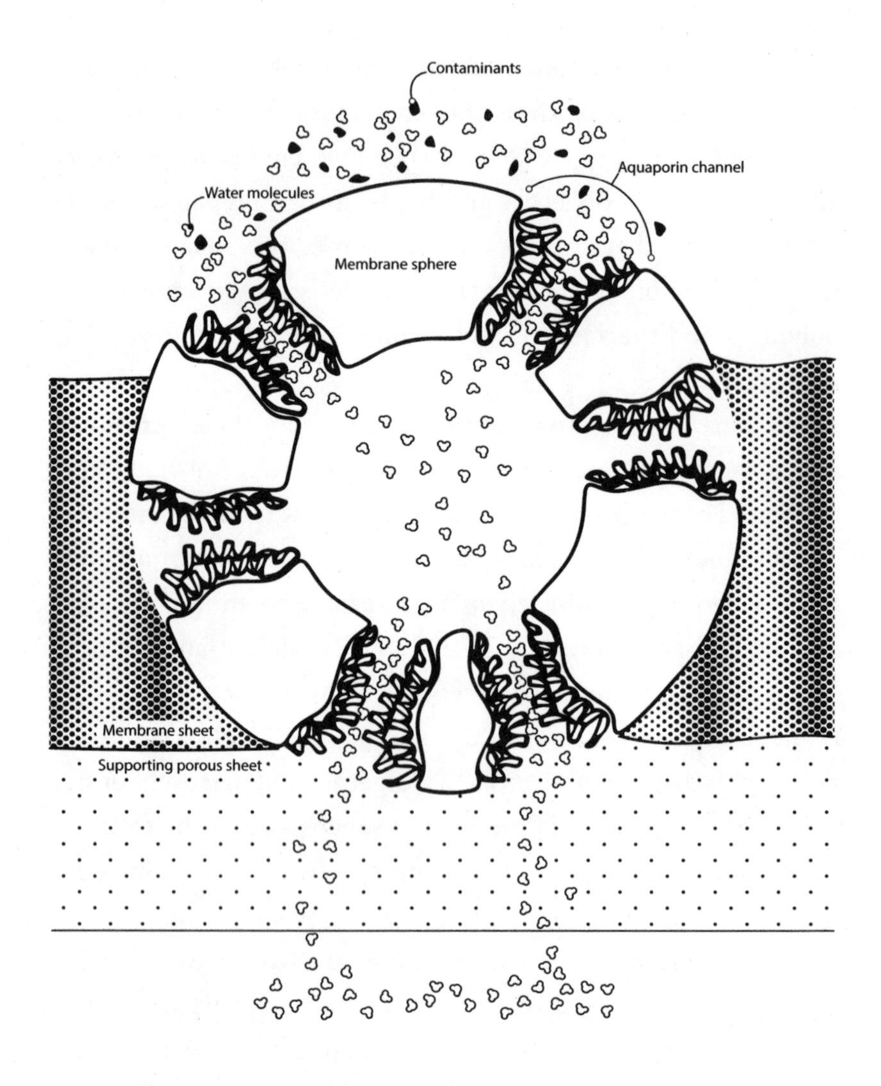

A cross section shows an aquaporin-bearing vesicle, with water traveling from one side of an aquaporin filter to the other. Water from a contaminated solution (at the top) flows through one aquaporin channel into the vesicle and then flows out of the vesicle through a second aquaporin channel. The vesicle sits in a membrane sheet, which is strengthened by a supporting porous filter sheet. After water (and only water) travels through the aquaporin channels (in the vesicle), it passes through the supporting sheet, yielding a pure water solution (at the bottom).

a vesicle sheet that was cheaper and stronger than a flattened sheet, and they demonstrated that the transit of water through the vesicles, which required each water molecule to pass through two aquaporin channels, didn't slow water flow too much. The relative ease of producing vesicle sheets compared to flattened sheets more than made up for the slight reduction in efficiency from requiring water to make two channel passages.

As long as a cell membrane effectively separates different watery domains from one another (the inside of a cell from the outside), it is doing its job. However, a cell membrane doesn't have much structural integrity: it's just a very thin fatty layer that forms the shell of a cell. It can't withstand the kind of forces required for a functional water-filtering membrane. A single, flat cell membrane is less than 10 nanometers thick, and even the beautifully designed and engineered vesicle sheet proposed by Hélix-Nielsen and his colleagues, which is about 200 nanometers thick, is still thinner than an oil slick on a puddle of water. The aquaporin vesicle sheets needed to be supported by a much stronger structure.

To solve this problem, the Aquaporin A/S team devised a method to layer the aquaporin vesicle sheet on top of a porous material. Hélix-Nielsen likens the structure to a sponge cake covered in a thin layer of frosting that is studded with raisins. The cake layer corresponds to the supporting porous material, the frosting represents the vesicle sheet, and the raisins are the embedded aquaporin-containing spheres. It's an exquisitely engineered structure, and Aquaporin A/S has figured out how to build it on an industrial scale.

The dream of aquaporin-based water purification has already proven itself in a high-profile test: in 2015, Danish astronauts used Aquaporin A/S membranes to filter the water they drank in space, where water reuse is a critical part of any successful mission. Now, in partnership with a Chinese joint venture called Aquapoten, Aquaporin A/S is working to put faucet-based water filtration systems on the market soon. Holme Jensen and Hélix-Nielsen showed me a prototype during my visit and explained how it would work. A set of three or four filter cylinders, each about a foot long and a few inches in diameter, would sit in a small plastic cabinet under the sink. As water flows up into the sink faucet, it gets forced through the aquaporin membranes at double the rate of flow common for standard faucet-based filtration devices. Depending on the quality of water in the water supply, the Aquaporin A/S filter would need replacement every half year or so.

o

Beyond sink-based water filters, Hélix-Nielsen has thought about many ways to optimize water use. He points out that most people use the same quality of water for all their water needs. In the United States that means that people use the same water source to fill drinking glasses, clean clothes and dishes, water gardens, and flush toilets. In many parts of the developing world, similarly, a single contaminated source of water is used for eating, drinking, laundry, and irrigation. Holme Jensen, Hélix-Nielsen, and Aquaporin A/S want to change that by dramatically expanding the use of water that is "fit for purpose." This idea has begun to catch on. A building at MIT, for example, has already adopted the approach by installing dual water systems: one that provides ultraclean water for drinking and dishwashing, and a second that provides a separate stream of recycled water for toilets and irrigation.

If the home water filters succeed, Aquaporin A/S has plans to pursue a new set of technologies that will change how we use water in the twenty-first century. Hélix-Nielsen imagines reducing agricultural water use and water waste around the world with an aquaporin-based forward osmosis system—a goal with potentially revolutionary consequences, given that roughly 70 percent of the planet's freshwater used today goes into agriculture. He showed me a prototype of the very large filter cylinder that the system would use and then explained how the system would work. Highly concentrated fertilizer would flow along one side of an aquaporin filter, and along the other side would flow the kind of runoff water that today farms just allow to wash away. Because the concentration of the fertilizer solution would be higher than the runoff water, osmotic pressure would draw water from the runoff water side through the aquaporin filter, in effect, drawing purified clean water from the dirty runoff water to dilute the fertilizer. This process would have the dual advantage of, first, reducing the volume of the runoff water and, second, reducing the amount of clean water needed to dilute the fertilizer—a win-win for water use. The same kind of system, he explained, could be used in a laundry facility. After being used to wash clothes, dirty water would pass on one side of an aquaporin-equipped filter that had a highly concentrated detergent on the other side. The concentration imbalance between the detergent and the runoff water would draw water (purified from its transit through aquaporin channels) from the dirty side and dilute the detergent. The runoff water volume would be reduced, and the diluted detergent would then be used in subsequent laundry cycles, increasing water reuse.

The brilliant biological discoveries of scientists like Peter Agre and the engineering innovations of people like Peter Holme Jensen and Claus Hélix-Nielsen are evidence that we are on the verge of transforming how we purify water and how we design our water systems. We are living through a revolutionary moment, one that Hélix-Nielsen likens to the moment when the automobile entered mass production. Just as the Ford Motor Company did a century ago, Aquaporin A/S aims to scale up a relatively new technology to deliver it economically and abundantly to millions—even billions—of people. "I think of our company like Henry Ford's," he said. "Ford didn't invent the car, but he built car production to scale, proved the technology, and delivered it to the masses."

4

CANCER-FIGHTING NANOPARTICLES

In 1971, the United States launched the War on Cancer, projecting an eight-year effort with a cost of $100 million. Today, more than forty-five years and $100 billion later, we can claim success in diagnosing and treating some cancers, but we are far from having won the war. Almost 600,000 Americans and more than 8 million people worldwide still die of cancer each year.

The War on Cancer's ambitious goals were based, with good reason, on a remarkable set of new insights into biological processes. By 1971, the molecular biology revolution had given scientists new ways of understanding disease. As we've seen, that scientific revolution had revealed biology's parts list—namely, the DNA, the RNA, and the proteins of viruses, bacteria, and the cells of complex organisms. The components and mechanisms of living things were being revealed at a stunning new level of resolution. And these new insights set the stage for new approaches to medical interventions including vaccines, drugs, diagnostic tests, and more. The next step seemed obvious. Why not use this new understanding of molecular biology to defeat cancer, one of humankind's most terrifying medical problems?

In retrospect, the misplaced confidence that an eight-year, $100-million effort could change the game in the fight against cancer rested on a limited appreciation of the complexity of the disease. A breakthrough discovery in 1970, for example, identified a gene from the Rous Sarcoma Virus (a chicken-infecting virus) that could convert a normal cell into a cancer cell and provided a window into possible cancer mechanisms. When Rous Sarcoma Virus (RSV) infects a chicken, the chicken's cells may incorporate a copy of a gene from the virus, called an oncogene. That virus-derived oncogene can disrupt the cell's normal processes, turning a normal cell into a cancer cell. A torrent of subsequent experiments showed that, in addition to cancer-causing genes that arise from extrinsic sources, like viruses, cancer can also arise from intrinsic sources, predominantly inherited genes, which can be activated by gene mutations. These cellular "proto-oncogenes" are usually silent, but they can be activated by mutations arising from any of several factors: by DNA-breaking events, like radiation or smoking; by viral infections that integrate foreign DNA into cellular DNA; or simply by errors that occur in DNA during cell division.

During normal development, cells divide and mature in carefully orchestrated sequences. Normal cells control their number, their function, and their location with exquisite precision. They follow a program to develop into their mature state—as a skin cell, say, or a liver cell or a lung cell. In doing so, they control their rate of division and their maturational pathway, in part, by detecting and destroying cells that have developed mutations in their DNA.

One of cancer's deadly hallmarks is that its cells divide without control, unlike normal cells that stop dividing when they recognize that they've made enough lung cells or brain cells. Cancer cells also don't follow normal pathways of maturation. They don't respect normal constraints on their location: cancer that begins in one location will in time send its progeny to other locations, metastasizing to set up new cancer sites and significantly increasing the difficulty of treatment. And, perhaps most perniciously, cancer cells don't self-edit to kill cells with DNA mutations; instead, they accelerate mutation, producing new cell variants that can evade the body's defenses.

The discovery of oncogenes opened the possibility of curing cancer by, first, developing techniques to identify oncogenes, and then, designing drugs that could turn them off. This has proven to be a powerful and nearly miraculous strategy, but only against some cancers. For example, imatinib, a drug approved by the FDA in 2001 and known commercially as Gleevec, targets the protein product of a cancer gene in chronic myelogenous leukemia (CML), a cancer of blood cells. In CML, a gene mutation turns on a protein that promotes abnormal cell division. Gleevec blocks the action of that protein, and it has delivered lasting remissions for many patients. Gleevec has increased the survival rate for patients five years after their diagnosis with CML from only about 30 percent to over 80 percent.

Many other cancers remain incurable and deadly. The DNA of cancer cells constantly mutates, giving rise to new oncogenes and cellular mechanisms that use an almost inconceivable set of tactics to promote the disease and evade treatment. Because cancer cells lack the self-regulatory mechanisms that either correct mutations or kill cells that harbor them, mutations accumulate as they divide, producing variants of the parent cells that can sometimes survive conditions that the parent cells cannot. For example, if an anticancer drug targeted to a specific cancer protein kills an original cancer cell and all of its genetically identical progeny, a drug-resistant variant often survives and propagates a new population of drug-resistant cancer cells.

The most effective strategy to fight cancer is to prevent it, of course. And since 1971, we've made considerable progress on that front, in large part by identifying and reducing our exposure to carcinogens, among them asbestos, radiation, and a variety of chemicals. But despite what we've learned, we still expose ourselves to dangerous carcinogens—by smoking cigarettes, by basking unprotected in the sun, or by refusing vaccines against viral infections. According to some estimates, more than a third of all cancers today are preventable.

In a way, that's good news. With more effective anti-tobacco campaigns, better sun protection, and new antiviral vaccines, we should be able to reduce cancer rates significantly. But even if nobody in the world smoked, if beachgoers everywhere wore sunblock, and if all children received vaccination against hepatitis and human papilloma viruses, cancer would still endanger millions of lives each year. We still don't know the causes of many cancers, so prevention alone cannot eliminate the threat they pose to us.

The next best strategy in the fight against cancer is early diagnosis followed by effective treatment. Here, too, we have made impressive progress. In recent decades, we have developed powerful new imaging techniques and blood-based tests that can detect cancer at earlier stages than before. Using mammography and colonoscopy screenings, for example, doctors today can identify breast and colon cancer much sooner than they could in 1971. And these screenings in combination with follow-up surgery, chemotherapy, or radiation have significantly increased patient survival. In the past forty years, the five-year survival rate for breast cancer patients has increased from 75 percent to more than 90 percent and for colon cancer patients from under 50 percent to more than 65 percent.

These technological advances represent remarkable victories won by determined researchers, clinicians, and patients. But most cancers are still not detected early enough. Although various cancers behave differently, current standard imaging techniques typically can only detect cell masses once they have grown to several millimeters or even centimeters in diameter, and even then the images don't confirm whether a cell mass is cancerous or benign. Making that determination requires invasive biopsies and more testing. All of that costs money, takes time, and allows cancer cells to continue to grow.

Blood-based detection tests face the same challenges. These tests look for biological traces, or "signals," left by cancer cells in the blood. They do a good job of picking up the signals of prostate and ovarian cancer, for example. But as with imaging, blood-based tests can only detect signals relatively late in a cancer's progression, and they can't always discriminate clearly between cancer and noncancer signals. Making definitive diagnoses, once again, requires invasive procedures, more money, more testing, and more time. And in the fight against cancer, time is of the essence. Early detection can make the difference between survival and death.

The imaging and blood-based diagnostic tools at our disposal are vastly more sensitive than a decade or two ago, but they both have similar limitations. They lack the sensitivity to detect cancer early in its outset when it's only a tiny mass of cells; they aren't selective enough to quickly determine whether the masses they can detect are cancerous or benign, so they often require invasive procedures to confirm a cancer diagnosis. They are also expensive. Together, these factors seriously impede the ability to combat cancer most effectively, and millions of people still die each year as a result.

To raise the fight against cancer to the next level, we need more accurate, faster, safer, and cheaper ways of detecting the disease early. Thanks to a recent discovery made by Sangeeta Bhatia, a biological engineer and physician by training who has become a pioneer in the fascinating world of medical nanoparticle technology, we may well soon have one. Bhatia has devised a urine-based test that promises to detect cancers remarkably early—that is, when cell masses are as much as twenty times smaller than the smallest masses that today's best imaging techniques can detect. The test, she hopes, will soon be as fast, reliable, and cheap as the over-the-counter pregnancy tests now available in pharmacies everywhere.

The idea may sound far-fetched. But it's not.

o

Even as a graduate student, Bhatia was a rising star. Before she had completed her PhD and MD programs, at MIT and Harvard, she received her first faculty appointment, at the University of California, San Diego, in 1999. And in 2005, after hearing about the fascinating ways in which she was bringing together medicine and engineering, we succeeded in recruiting her back to MIT. She continued her pioneering work, and in 2007 when we started to identify engineers who could work with cancer biologists at the new Koch Institute for Integrative Cancer Research, she became an obvious candidate.

To learn more about her work, one afternoon I stopped by her office. Bhatia radiates a calm that I warmed to immediately. But as we talked, I quickly realized that behind her unassuming, gentle demeanor was a mind running at a blazing speed. I was astonished at the pace, the scope, and the intensity of her work, which boldly brings together biology, medicine and engineering in ingenious new ways.

Now, more than a decade later, I'm still astonished. Like Angie Belcher and the other pioneers in this book, Bhatia moves easily across disciplines. In her graduate work she used tools from computer-chip manufacturing to design artificial human organs. At UC San Diego, she made a name for herself as a nanotechnologist with a knack for solving biomedical problems. And today, at MIT, she works in a dizzying variety of capacities. Her multidisciplinary interests are reflected in her faculty appointments in several MIT departments and also at Brigham and Women's Hospital.

Throughout her career, Bhatia has focused on very tiny technologies, and now as director of the Marble Center for Cancer Nanomedicine, she designs novel platforms to understand, diagnose, and treat human disease. Her commitment to making a difference not just in science but also in the lives of patients has led her and her trainees to found several biotechnology companies, each of which occupies a place, in her words, "at the intersection of medicine and miniaturization."

Miniaturization for Bhatia means working with matter at an extremely small scale—the nanoscale. The particles she works with range from 5 to 500 nanometers in size. For an idea of just how small that is, consider this: if you were working with 10-nanometer particles, you'd need 100,000 of them to fit across the width of the 1-millimeter-diameter period at the end of this sentence.

Nanoparticles are simply very tiny pieces of matter. They come in dozens of shapes (spheres, rods, pyramids, dodecahedrons) and can be made of elements (silicon, iron, gold) or biological materials (proteins, DNA strands). They can be made to have specific sizes and compositions to suit specific purposes. For example, iron oxide is a very useful material for magnetic resonance imaging (MRI), but its extreme reactivity with water and even with the humidity in the air makes it difficult to use in biological and medical applications. However, iron oxide nanoparticles can be stabilized against reacting with water by coating them with a layer of sugars, polymers, lipids, or metals. The choice of coating materials can change the function of a nanoparticle to suit a particular application.

Curiously, a material at the nanoscale can have different properties from the same material at the macroscale. Gold nanoparticles, for example, don't appear gold. Instead, they appear red because they reflect red light. Two thousand years ago, Roman artisans unwittingly created gold nanoparticles in their glassmaking practices, which allowed them to produce highly valued ornate glassware that, because of those nanoparticles, took on a distinctive red tint. We also know now that simply changing the size of a nanoparticle can sometimes change its properties. Cadmium selenide naturally forms large black crystals, for example, but liquid suspensions of different sizes of cadmium selenide nanoparticles glow in different colors because light interacts differently with the compound at different sizes: two-nanometer particles glow blue, four-nanometer particles glow yellow, and seven-nanometer particles glow red.

By studying properties like these, scientists and engineers have learned how to control the size, composition, and architecture of nanoparticles with amazing precision. And they recognized that nanoparticles could be used for all sorts of surprisingly everyday purposes. Today we put silver nanoparticles into toothpaste because they kill bacteria. We put titanium oxide and zinc oxide nanoparticles into sunblock because they provide remarkable sun protection. We put nanometer-sized particles of a material called carbon black into car tires because they enhance rubber's traction and durability.

In recent decades, medical researchers have recognized that they, too, can put nanoparticles to use. Of particular interest to Sangeeta Bhatia and others is the newest generation of nanoparticles, which can be fabricated and coated in a myriad of ways to engineer a new generation of tools for biomedicine. It is these nanoparticles that ultimately made it possible for Bhatia to devise her groundbreaking cancer diagnostic—a wonderful story of how convergence-based research can lead to powerful new technologies.

This chapter of Bhatia's story begins in 2000, when she and her colleagues were exploring ways in which nanoparticles might ferry imaging material through the bloodstream to a specific tissue in the body—a process that would give doctors and researchers a clearer view of whether a disease process was underway in that tissue. At the same time, they were exploring how nanoparticles might be used to deliver drugs to tackle diseases in a specific tissue. If you suspected liver disease, for example, you could send special imaging nanoparticles to the liver, where they would reveal whether a disease process was underway; if signs of liver disease showed up, you could use other nanoparticles to send drugs right to where they were needed most. An ideal nanoparticle, carrying either imaging material or a drug, would attach selectively to its specific target while bypassing all other kinds of tissue.

This kind of complex, discipline-crossing work calls on a wide range of expertise, and Bhatia has enjoyed collaborating for more than fifteen years with two colleagues from her years at UC San Diego, Michael Sailor and Erkki Ruoslahti. Sailor, a chemist and materials scientist, has focused his research on determining how to build nanoparticles from new materials, such as porous silicon, one of the key materials used in semiconductors, and iron oxide, a material visible by MRI. Ruoslahti, now based at the Stanford-Burnham Institute and the University of California,

Santa Barbara, has long studied a family of cell-surface adhesive proteins that allow specific kinds of cells to come together and assemble into precise multicellular organizations.

Together, the three began investigating how they might tag Sailor's nanoparticles with sequences from Ruoslahti's sticky adhesion proteins so that they would then bind to specific tissue sites—the blood vessels of a tumor, as one example. The protein sequences would serve as "zip codes" to guide the nanoparticles through the bloodstream to the desired target-tissue "address." Once there, these adhesion proteins would bind to the tissue while still attached to the nanoparticles. If those nanoparticles had a composition that could be detected by MRI, then the target tissue, too, would become visible by MRI.

It was an alluring idea. Bhatia knew that iron oxide nanoparticles are particularly useful in MRI, so she and her colleagues tried attaching Ruoslahti's protein zip codes to Sailor's iron oxide nanoparticles. But here they encountered a problem. Because the nanoparticles were so small, they had to create clusters of nanoparticles at a critical density to make them visible to MRI, but those clusters turned out to be too big to travel in the blood through the small capillaries that could bring them to the target tissues.

Bhatia realized she needed a different strategy. She had to figure out how to deliver nanoparticles in a sufficient quantity and density to be visible to MRI but not create clusters so large that they would clog the capillaries through which the nanoparticles needed to travel. It was a tough problem, but eventually she had an idea: What if she could design nanoparticles that would cluster on demand at the target tissue? That is, what if she could fabricate nanoparticles that would travel individually through the bloodstream and then cluster when—and only when—they reached the target tissue? Rather than giving a nanoparticle a protein "zip code" that would deliver it to a particular tissue's "address," Bhatia imagined a new kind of label that would use features of the biology of the target tissue itself to cause nanoparticles to cluster only at the target tissue.

She and her team set to work. First, they made two sets of nanoparticles, with each set carrying one member of a pair of proteins that typically bind to one another with great avidity. When the two different sets of protein-labeled nanoparticles were exposed to one another, they would interlock and form large clusters. To prevent the proteins from interlocking and forming clusters as they traveled through the bloodstream, Bhatia's team tethered a "shield" to the nanoparticles to hide the binding proteins from one another. They used an inert substance (polyethylene glycol, or PEG) for the shield, and they tethered the shield to the nanoparticle using a short segment of a protein.

The PEG shields allowed the nanoparticles with their now hidden clustering proteins to travel through the bloodstream without interlocking and clogging up blood vessels. (The PEG shields had the added benefit of disguising the nanoparticles from the body's defense processes. Anything that travels through the bloodstream faces the body's defense mechanisms, which detect and remove foreign material from the body, and the PEG shield hid the nanoparticles from these mechanisms.)

The team now was ready to send nanoparticles through the vascular system, including its tiny capillaries, in search of a target tissue. They ran a series of tests, and their new design performed just as they'd hoped it would: their nanoparticles didn't form clusters in transit. Instead, they traveled easily through the bloodstream and evaded detection and removal by the body's defenses.

The next challenge was getting the particles to cluster when they reached the target tissue. With their shields on, the nanoparticles wouldn't cluster, so Bhatia and the team had to figure out how to remove the shields at—but only at—the target site. In grappling with this challenge, Bhatia had a breakthrough idea. What if she could piggyback the task of removing the shields on some tissue-specific activity? Pursuing that idea, she developed an ingenious strategy: she would attach the nanoparticles' shields with a protein that the target tissue would cut using its own tissue-specific enzymes.

Enzymes are proteins that cut other molecules with high selectivity. They come in thousands of varieties, each found in particular locations and with particular cutting targets. Enzymes work like molecular scissors, each cutting a specific molecule, called the enzyme's substrate. Many enzymes have proteins as their substrates and make their cuts at specific amino acid sequences in the target proteins. By cutting a protein at a specific amino acid site, enzymes serve critically important functions: they can turn inactive forms of proteins into active fragments or an active protein into inactive fragments. Among the remarkable features of enzymes is that when an enzyme cuts its substrate it does not inactivate itself. A single enzyme molecule can act again and again, cutting a large number of its specific substrate molecules,

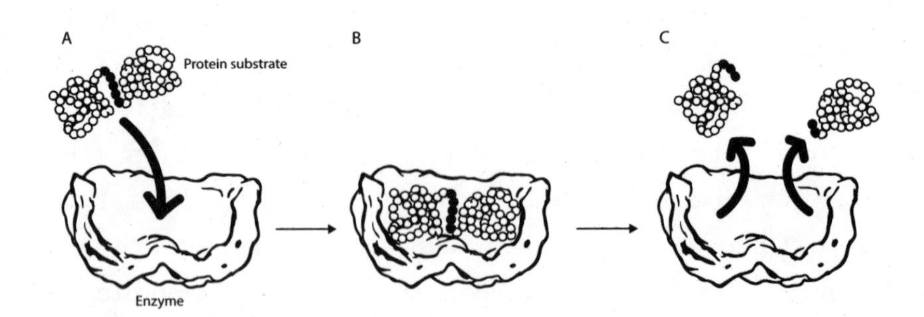

Enzymes cut proteins at specific sites. (A) An enzyme binds to its protein target (substrate). (B) The enzyme cuts the protein at the enzymatic site (dark beads). (C) The cut fragments are then released.

always at its specific cutting site. And it can do its work very rapidly. Some enzymes can cut as many as a thousand, or even ten thousand, substrate molecules a second.

The specificity and speed of enzyme action make possible some exquisitely selective and efficient biological processes, like blood clotting, food digestion, and cell movement in cancer metastases. Bhatia devised a way to harness enzyme action to cluster her nanoparticles. She began by identifying a tissue-specific enzyme and designed a protein fragment with that enzyme's cutting site. She then used that specific protein fragment to tether the PEG shield to the nanoparticle, reasoning that when the nanoparticles reached the tissue, the tissue's enzymes would recognize the tethering proteins and cut them. The cut would release the shield and expose the interlocking proteins they had been hiding. Once exposed, the interlocking proteins would find one another and cause the nanoparticles to cluster.

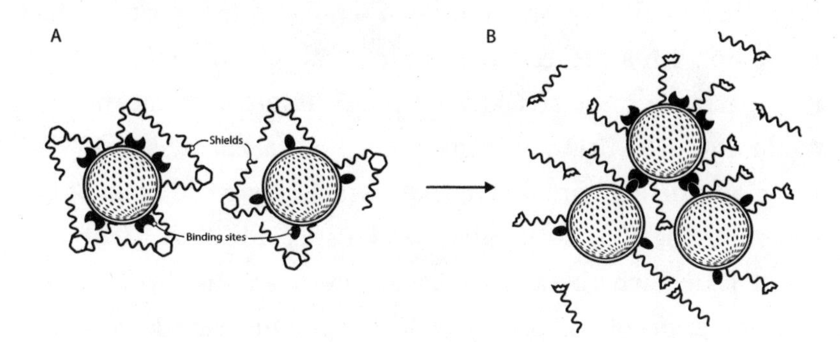

Enzyme-induced binding of nanoparticles. (A) Two different nanoparticles carry binding sites for one another. The binding sites are shielded to prevent interactions between them. Each shield contains a cutting site for an enzyme (hexagon). (B) The binding sites are revealed when an enzyme cuts at the enzymatic site (separated halves of hexagons). After an enzyme makes its cuts, the shields fall away to reveal the binding sites, enabling the nanoparticles to bind to one another and form clusters.

As they assembled these complex nanoparticles, Bhatia and her team had to be able to follow the components closely. To that end, they attached markers to confirm that the particles, the interlocking proteins, and the shields with their tethers had all assembled correctly. They used a fluorescent tag to follow the protein that tethered the nanoparticle to its PEG shield.

Complicated! Or, as my graduate advisor used to say when I would propose outlandishly complex, multilayered research strategies to him: "Experimental acrobatics, destined to fail." But in Bhatia's hands, the acrobatics worked beautifully. In 2006, she and her team reported the successful enzyme-mediated clustering of nanoparticles in cell cultures, and in 2009, they demonstrated tissue-specific delivery and visualization of nanoparticles in the spleen and bone marrow.

It was brilliant work. By devising an ingenious combination of interlocking proteins and removable protein shields that Bhatia calls "synthetic biomarkers," she and her team had come up with a successful and practical technique for getting nanoparticles to cluster at a targeted tissue. That meant they could now use MRI to peer inside the body at that tissue. If MRI revealed a disease process in that tissue, they could readapt the same nanoparticle strategy to send a specialized drug directly to the tissue, where nanoparticle clustering would bring a high concentration of the drug to address the problem locally in a precise, focused way.

Bhatia's technique had all sorts of applications for diagnosing and treating diseases in the body's different organs—for doctors who need to find tumors, for example, or monitor liver damage from a progressive liver disease. But Bhatia quickly recognized a less obvious application that had not been a focus of their work initially. With a bit of serendipity, she discovered that her new nanoparticle technology could provide a faster and more sensitive way to diagnose cancer and other diseases.

o

Doctors who find a single mass of cancer cells in one location often can use surgery or targeted irradiation to remove or kill it. But once the cancer has metastasized, the job gets a lot harder because the multiplying cancer cells take up residence in many new sites, where they are hard to find and treat.

Metastasis is one of cancer's most deadly features. To spread to new locations, cancer cells need to overcome the natural tissue and molecular barriers they encounter along the way. The body's organs are self-contained structures with their cells organized in highly precise architectures. Invading cancer cells must overcome an organ's molecular structures that keep normal cells in their correct places. To clear space in the tissue to permit their invasion, cancer cells turn on special enzymes that cut through the proteins and other molecules in their way.

Bhatia and her team realized that they might be able to adapt their new tissue-visualization technique to "light up" not just normal organs to monitor their response to disease but also to follow the cells of the diseases themselves, including developing cancers. If the shields they used to prevent clustering of their nanoparticles in transit through the bloodstream were tethered with fragments of proteins that cancer enzymes usually cut, wouldn't that mean that the tethers would be cut if—and only if—they encountered cancer cells? If so, then the previously shielded proteins would be exposed and would interlock, causing their attached nanoparticles to cluster at the site of a tumor. And that, in turn, would potentially make cancer visible to MRI far sooner than has previously been possible.

Bhatia and her team ran experiments to test their theory and succeeded: they could visualize experimental tumors on MRI. It was a triumph. And then, with a bit of luck and brilliant scientific insight, they made a discovery that would propel their work in a different and potentially even more important direction.

The discovery came as they were evaluating the results of their experiments. In doing so, they noticed a puzzling detail. In addition to seeing the hoped-for specific MRI signal at tumor sites in their laboratory mice, they also detected an unexpected fluorescent signal in the mice's bladders.

It would have been easy to dismiss a signal in the bladder as an experimental artifact not worth noting. The bladder accumulates urine, after all, and urine is a waste stream that might naturally contain elements of the protein shields and tethers that the kidney had filtered out as it routinely processed the blood. But there could be other explanations, and the unexpected finding stumped Bhatia's students, who feared that the aberrant signal in the bladder indicated that the experiment had failed. But when they brought their result to Bhatia, she recognized that something unusual and interesting was happening. "My medical training told me that there was no way that our fully assembled nanoparticles could make it into the urine," she said. "They're just way too big to pass through the kidney's filter."

Bhatia felt compelled to understand what was going on, and with some sophisticated biological detective work, she figured it out. The fluorescent tags generating the signal in the bladder were not attached to intact nanoparticles. Instead, they had made it through the kidney on small pieces of the protein tethers that had attached the PEG shields to the nanoparticles. Bhatia's team had tagged the tethers so that they could follow them through the complex steps in the experiment—first to verify that the

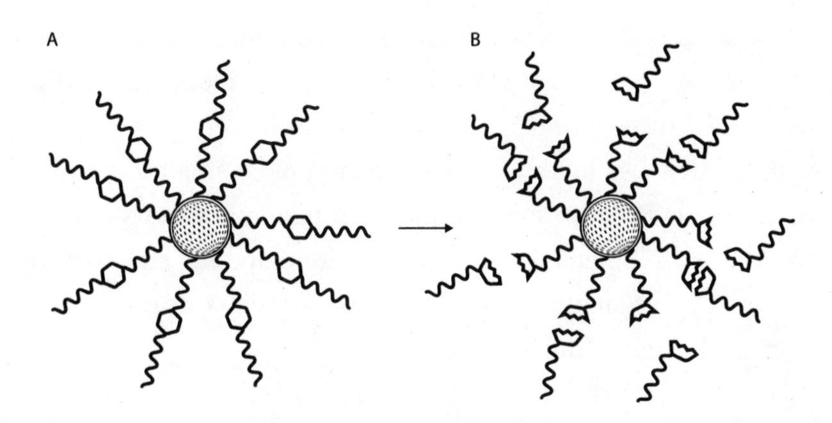

Diagnostic nanoparticles. (A) A nanoparticle decorated with proteins that carry a cutting site for a disease-specific enzyme (hexagon). (B) When a disease and its enzyme are present, the enzyme cuts the protein and releases the distal protein fragments, which travel through the blood and are filtered by the kidney into urine. The small fragments are detected in the urine.

tethers and the shields had attached to the nanoparticles, and then to keep track of them as they traveled through the bloodstream to the tumor. What they hadn't realized, however, was that when the cancer enzymes cut the protein tether holding the PEG shield, the released fragments were tiny enough to pass through the kidney's filter into the urine where, still fluorescent, they accumulated at a high enough density to be visible.

As soon as Bhatia understood what was happening, she recognized that she and her team had discovered a biological mechanism that they could potentially deploy as a simple and reliable means for very early cancer diagnosis. If so, it would be a revolutionary clinical advance.

In the years since making that original discovery, Bhatia has streamlined her diagnostic test so that it now works much like an over-the-counter pregnancy test that is noninvasive and inexpensive with a very high sensitivity. Its sensitivity is truly remarkable. Recall that today's imaging and blood-based tests can detect tumors only as small as about one centimeter. Bhatia believes that her method could detect tumors smaller than five millimeters. Because some tumors can take many years to grow from five millimeters to one centimeter, finding and treating tumors at this smaller size would give doctors a dramatic head start in the fight against cancer.

To accelerate the clinical development of this remarkable new technology, Bhatia has launched a company called Glympse Bio, which anticipates the development of a market-ready urine test by 2020. She and her colleagues at the company plan to develop a set of new diagnostic technologies that will enable the early detection of not just cancer but many other types of disease as well. "We spend so much time and money trying to keep our patients from dying during the last stages of a disease," she told me. "But for many diseases, if we can detect it early, we can treat it before it has become untreatable."

As Bhatia's work makes clear, nanotechnology promises to revolutionize medicine by dramatically improving detection and therapeutics and reducing costs. Bhatia has many committed colleagues in her quest. For example, nanoparticles that release their drug payloads slowly over time currently provide long-lasting treatment from a single injection for cancer and other diseases, and nanoparticles carrying a variety of imaging agents enhance the power of medical ultrasound and MRI. The accelerating convergence of nanoengineering and biology in the coming decades will bring new technologies that we can't even imagine today—nanotechnologies that will profoundly transform how we deliver health care and battle disease. Bhatia has no doubts at all about this. "The future," she says, "is small."

5

AMPLIFYING THE BRAIN

Jim Ewing returned from a family vacation in the Cayman Islands a changed man. An engineer at a climbing rope company, Ewing had been an avid rock climber since his teenage years. He was scaling the cliffs overlooking the ocean when a rope malfunction sent him plummeting fifty feet onto a sharp rock crag below. Over the next year, he recovered well enough from his broken pelvis, dislocated wrist, and bruised lung, but despite multiple surgeries and physical therapy, his left leg, badly mangled from the impact, still gave him insufferable pain and prevented him from engaging in even the most benign activities of daily living. He could not even put on his socks without excruciating pain, and he had given up hope of returning to any semblance of his former, active life. He faced a tough decision: Should he keep or give up on his natural leg? Should he undergo an amputation and begin a new life dependent on prosthetic devices? The decision terrified him.

As an engineer, Ewing continued to search for possible medical solutions but without much hope. He recalled a climbing buddy from decades ago who—on prosthetic legs—had accompanied him scaling some of the most challenging rock faces in New England. Ewing decided to reach out to him.

Ewing's buddy, Hugh Herr, now a professor, directs the Biomechatronics Group at MIT's Media Lab. Herr offered a very different view of Ewing's future. Herr works at the forefront of the design of smart prosthetics and has a profoundly vested interest in the field: in 1982, at age seventeen, he lost both of his lower legs in a mountaineering accident. After his injury, he resolved to climb again, but the rudimentary prostheses of the time made that ambition an impossible dream. They supplied none of the assistive forces that we get from joints and muscles, and they certainly didn't communicate with the nervous system. They could provide him with sufficient stability to keep him upright and mobile but not nearly enough agility to climb a mountain face.

Herr didn't let go of his dream, however. Instead, he took matters into his own hands. While still in school, he began to make prosthetic legs that allowed him to climb again and, for the first time, he got serious about his studies. He finished high school and then studied physics at a local college. Herr did well enough there to get into graduate programs at MIT and then Harvard, studying mechanical engineering and biophysics, and gained a deep understanding of mechanics and mathematical modeling of motion and force. He returned to MIT for his postdoctoral work, where he embarked on designing and building a revolutionary new generation of computer-assisted prostheses. Today, using prosthetic legs he designed himself, he's climbing again, avidly and expertly, and he navigates the world so gracefully that no one would imagine his legs are not biological. Herr has become one of the leading lights in imagining future technologies that might return people with amputations or paralysis to fully mobile lives.

Herr got Ewing's call at an opportune moment. He had been working on a project to design a new kind of device that together with a novel amputation method would enable the recipient to interact with a robotic prosthesis. Ewing was still torn between his options.

The computers and mechanics in the ankles that Herr designs simulate with remarkable accuracy the behavior of the human ankle. The movements of an ankle or a hand, or any of our actions, are the products of a complex motor system. That system engages the nervous system and muscles to translate our intentions—for example, the decision to walk up a flight of stairs—into a motor program that carries out remarkably complex tasks with a high degree of accuracy and very little active thinking. We have an acute awareness of some of the components of that system, while other components function essentially autonomously, well below our normal level of awareness.

Take a very simple action: when you sit with your right leg crossed over your left leg, if you want to flex or extend your dangling right foot, you do so by activating the muscles in your calf. To accomplish this seemingly simple task, you don't micromanage all the neurological and muscular details to carry out that action. To point your foot down, you don't have to direct the muscles in the back of your calf to contract, and you don't have to direct the muscles in the front of your calf to relax, so as not to resist that action. While you're conscious of your intent, which initiates the action, you're unconscious of the many steps in between that accomplish your intent.

If I've persuaded you to cross your legs and carry out this little demonstration, the appropriate muscles to raise or lower your foot have been activated by signals arising from nerve cells in your spinal cord. The nerve cells that directly activate muscles are called motor neurons, which sit in the spinal cord, a long column of the nervous system that extends from the brain down the center of the back. The cell bodies of motor neurons, essentially their "base of operations," contain the cell nucleus (with its DNA) and most of the apparatus that supports a cell's activities (for example, reading DNA into RNA, and RNA into proteins). Each motor neuron has a long, thin branch, called an axon, which extends out of the spinal cord. Thousands of motor neuron axons bundle together as nerves that reach specific muscles and make connections that drive muscle activity.

In addition to motor neurons, the spinal cord contains myriad other neurons that interact with one another to form circuits that orchestrate our movements. Motor neurons drive muscle contractions, while sensory nerve endings in muscles and joint tendons send information about muscle activity—for example, whether a muscle is contracted or stretched—back to the spinal cord. The signals from the spinal cord to the muscles and from the muscles back to the spinal cord are carefully balanced to achieve the desired action. The coordination of the details of moving your ankles, such as the reciprocal activation of the muscles that point the foot down and the inhibition of the muscles that would point the foot up, takes place within the spinal cord. And much of that coordination happens responsively, adjusting to the position and movement of the ankle itself.

When a limb is lost, the connections from the spinal cord can no longer drive any activity. A prosthetic ankle can't respond to your intent. However, Herr has programmed the computer in his prosthetic ankle to mimic a normal ankle's movements, in effect by reproducing the kind of feedback that the brain and spinal cord exchange as someone alters his or her gait or responds to irregularities in a walking surface, like an incline in a path. To design the ankles, Herr studied the biology of walking by using advanced sensing technologies to follow ankle and leg movements. He also measured the energy a person uses in walking—both on biological limbs and on passive prosthetic limbs—and demonstrated the dramatic difference between the two. He measured how much force a normal ankle exerts during the "push off" phase of walking, when the trailing foot leaves the pavement to swing forward. Rigid prosthetic legs, ankles, and feet can provide stability, but they do not provide any additional momentum, and people using them have to invest enormous effort to swing them into place for each next forward step.

Herr modeled these biological factors into a computerized ankle, the latest version of which he calls "EmPower." Walking with EmPower requires no greater energy exertion than walking with a biological ankle. This is a remarkable achievement and a transformational advance for people like Herr himself, not to mention veterans returning to life and work following battlefield injuries. As good as it is for walking, however, the EmPower ankle doesn't allow users to, say, idly tap their feet along with a tune. This kind of higher-order problem is one of many that Herr is working on at his lab for Biomechatronics. Eager to learn more, I stopped by for a visit.

The lab is small, a two-story workroom, crammed full of equipment and "spare parts." When I arrived, I walked past lab benches filled with devices and tools, prosthetic limbs of various designs, and treadmills equipped with large banks of computers, each programmed to very precisely measure the pace, effort, and angular position of each joint of a leg during walking or running. I ascended a spiral staircase in one corner of the workroom to an upper set of offices, an architectural detail to maximize the lab's capacity. Herr's small, spare office contained only a table and chairs, and a few prosthetic legs that leaned up against the walls.

Herr has committed his lab and his life to inventing the future of a whole repertoire of assistive devices: computer-designed sockets that connect a device to the residual limb; prostheses with movements that mimic natural limbs; devices that detect signals in muscles so the wearer can move a prosthetic limb by intent; and brain-computer interface technologies that will, one day, connect the nervous system to a prosthesis, allowing the wearer to both move the limb and to feel its movement. Herr aims to design prosthetic arms and legs driven by the nervous system that will restore full functionality.

The ankles that Herr has already designed mark a triumph for prosthetic technology. With their new limbs, people who have lost a leg can now do the things they loved to do before their injuries—for example, stroll down a street or climb a rugged path and more. Importantly the EmPower ankle, unlike passive ankle prostheses, carries the same workload as a biological ankle. This technology represents a potentially transformational advance for everybody who has lost a limb. But, as is so often the case, these new technologies will not reach their transformative potential until we've figured out how to take them out of the lab and get them to market, which requires making them lighter, more adaptable, less expensive, and easier to maintain. To understand how innovations move into the marketplace, I paid a visit to one of the companies doing pioneering work on prostheses—namely, Össur, which is based in Iceland.

o

I flew to Reykjavik, Iceland's capital, in late October and checked into a hotel in the center of town. At 9:00 the next morning, I left the hotel for the Össur offices during a very long dawn, a reminder of how close we were to the Arctic Circle. When I entered the Össur building a half-hour later, the sun was just rising in earnest, casting low rays into the company's lobby, intensely illuminating the mission statement posted there on the wall: "Life without Limitations," a fitting prelude to what I was about to witness.

After making a call to say that I had arrived, the receptionist invited me to wait in a comfortable lounge. Photographs and video screens situated all around the room showed adults and children putting Össur prostheses to use: cycling, climbing mountains, playing running games, and just carrying out activities of daily living, like climbing stairs. For someone without an arm or a leg, these routine daily activities represent heroic achievements. I found myself drawn to a photograph of a beaming bride in a beautiful wedding gown who walked joyfully alongside her soon-to-be-husband. As any bride might, she held up the hem of her gown—revealing two prosthetic running blades. On a nearby video screen, a gang of gleeful children in brightly colored running garb raced across an athletic field, some sporting one prosthetic leg and others sporting two.

I had only barely enough time to begin to comprehend the new lives that smart prostheses can give back when two of Össur's leaders, Hildur Einarsdottir and Kim De Roy, greeted me in the waiting room. Over the next several hours they introduced me to the company and its products, including many of the most advanced computer-assisted prostheses available to amputees today. De Roy set a fast pace in both walk and talk as we crossed a long glass-walled corridor, which offered a beautiful view of a volcanic mountain range, and made our way to Össur's research lab for bionics.

Einarsdottir pointed out a set of computer-driven knees—

prototypes of what the company calls its RHEO KNEE®, which is designed to anticipate the wearer's motion. The RHEO KNEE® assists in walking and running, rather than simply enabling them, as rudimentary mechanical prostheses inevitably do. I was struck by the similarity to Hugh Herr's computer-animated ankles and learned that this was no coincidence. My hosts explained that the technology had originated in Herr's lab and that Össur had acquired the company that had first licensed it.

Össur now makes its knee for people around the world. Like the EmPower ankle, it anticipates the wearer's movements, using small computers housed in the RHEO KNEE® to mimic the kind of neural processing that would normally occur within the spinal cord. It provides sophisticated responses as the wearer navigates different conditions: climbing or descending stairs or a rocky mountain path; walking at a slow or a fast pace; squatting or standing up from a chair. The knee's embedded computers and electronics allow it to "think for itself," giving the wearer greatly improved agility over a passive joint, the kind of enhanced mobility that Herr's EmPower ankle also provides.

Putting its products into the marketplace requires that Össur's replacement feet, ankles, legs, and knees must be able, literally, to stand up to intense use, weight-bearing activity, and twists and turns in motion. And in some cases they must last a lifetime, because many amputees' medical coverage extends to only a single prosthesis and may not cover service or replacement components. Össur has developed the RHEO KNEE® into a long-lasting, highly responsive, regulatory-approved device without mandatory service requirements. To make all this possible, beyond the technology of the knee itself, Össur has designed technologies to manufacture the knees reliably and efficiently. The production of prostheses begins with brilliant design and finishes with expert manufacturing.

To move designs into production, the Össur complex includes a manufacturing facility. De Roy next led us into the Assembly Lab, essentially an on-site factory, to show me how the company built a manufacturing process to meet these demands. Össur uses the highest-quality metals (for example, the kinds of aluminum and titanium used for jet-engine parts) and state-of-the-art carbon fiber in manufacturing processes modified and monitored to achieve the precision demanded. I watched the components move through more than a dozen different machines that shape, hone, and polish blocks of metal and strips of carbon fiber into ankles, feet, and knees. Some of the components have an error tolerance of only eight micrometers, and they must retain this exquisite level of precision and strength reliably all day, every day, as the wearers navigate the constant challenges of walking, climbing, stopping, and standing through their daily activities.

From the Assembly Lab, De Roy briskly descended a couple of flights of stairs into a large, sunlit lobby that looked like an exercise studio with ramps, stairs, smooth and rough walking surfaces, stationary bicycles, and an array of other equipment. It was Össur's Gait Lab. Much of the equipment was in use, with people walking, jumping, and cycling, and I realized that, after spending a few hours at Össur, I no longer noticed that most of the people monitoring and using the equipment wore prosthetic devices. De Roy described some of the challenges designers had to solve to make it possible for someone wearing a prosthetic leg to walk easily up and down stairs. To demonstrate, he raised his trouser leg, pulled down his (very fashionable) sock, and showed me how his prosthetic ankle raises the front of his foot (which would have been his toes) just enough to ascend stairs with the same tempo that I do. By then, over the course of several hours touring the company together, we had walked close to a mile and navigated several flights of stairs, but I had not noticed any gait anomaly that would have led me to imagine that De Roy's left foot was prosthetic and not biological.

De Roy's ankle and foot are one of the company's newest products, known as the Pro-Flex, and he and Einarsdottir could not contain their excitement about its innovative design. The Pro-Flex uses entirely mechanical properties, with state-of-the-art carbon fiber and intricately designed junctions that provide mechanical push off and ankle rotation that give De Roy the versatility, balance, and power he needs, without a computer or a motor. The Pro-Flex comes much closer to replicating the ankle's natural motion than previous devices, and it does so in a lighter, lower-cost, and sturdier form than more complex prostheses.

o

The new materials, new computers, and new devices on display and in production at Össur have opened new possibilities for prostheses. They offer more versatility, balance, and power than ever before. But wearers of even the most sophisticated, computer-controlled, motorized knees still sometimes express a powerful frustration: they can't move their prosthesis as simply as they would move their natural arms and legs—that is, with intent. Instead, they often have the unsettling sense that "the knee is walking me."

Designing limbs that engage the nervous system and respond to the wearer's intent is the next great challenge in the world of smart prostheses. Össur is one of the companies inventing ways to meet that challenge, as I learned when I talked with Magnus Oddsson, the company's vice president of prosthesis research and development.

Oddsson radiates a kind of intensity that reminds me of my university colleagues, with a mind that is constantly at work several levels beyond our conversation. He speaks sparingly and with precision. In its work on intent control for prostheses, he explained to me, Össur's goal is to produce "clinically validated innovative solutions," meaning that the devices had to work in demonstrable, measurable ways for real people in real situations. They needed to improve motility and be embraced by the wearer. Beyond the device design, they had to provide medical evidence of value creation and cost savings, along with an enhanced capability for the wearer. For the company to succeed, a balance between state-of-the-art research innovations and market acceptability was essential. The devices not only had to work, but they also had to be recognized by established health care providers, and they had to last. To achieve all of that, Oddsson told me, "Simplicity is key."

Oddsson explained that Össur's strategy was to build on the wearer's own biology and not to try to re-create or reinvent the neuromuscular system. Össur's goal is to help people recover lost function, not to create new functions or superhuman abilities. As a result, the company has focused its efforts on the biological processes normally involved locally in the movement and control of muscles—in those of the leg, for example.

Össur's strategy, to enable "intent" to move a prosthetic ankle and foot, calls on the functioning elements of the neuromuscular system that remain after amputation. Recall that the motor neurons in the spinal cord have long axons that course through the limb to activate the muscles involved in a particular motion. Amputations leave the nervous system intact up to the point of the amputation, and most amputations below the knee preserve some of the muscles of the lower leg that would normally have raised or lowered the foot at the ankle joint. Recognizing this, Össur has developed biocompatible, wireless electrodes that sense muscle movement. Össur implants "myoelectric" sensors in the muscles that normally would control extension or flexion of the ankle (toes pointing down or up) and that respond to the state of contraction or relaxation of the muscle. The muscle-embedded sensors send their signals to a receiver located in the cuff of a prosthetic lower leg. The receiver then relays that information to computerized motors in a prosthetic ankle that extends or flexes the ankle to raise or lower the prosthetic foot.

Oddson illustrated Össur's progress with a video of "Kali," an amputee who is helping Össur test a prototype of this new generation of mind-driven prostheses. In the video, Kali wears a prosthesis attached to his right leg below his knee. Sitting with his prosthetic leg crossed over his natural leg, he rivets his attention on his foot. Then by simply willing it to happen, as he might have with his natural leg, Kali flexes and extends his foot, with his robotic ankle doing the work that his natural ankle would have done. After a few cycles of flexing and extending the prosthetic foot, he looks up with a smile of triumph that conveys a mix of delight, amazement, and pride.

Kali moved his ankle and foot by doing more than just thinking, of course. Despite missing his natural ankle and foot, his amputation preserved some of the muscles remaining in his leg below his knee, which allows him to be able to contract and relax those muscles. Before he received the myoelectric sensors that drive his computerized ankle, he had no reason to contract or to relax his lower leg muscles. Doing so would have served no purpose. But with a prosthetic limb that can receive signals from those muscles, now he can once again will, or express the intent, to move his foot. As this sort of technology becomes more common, surgeons performing amputations will change how they work; anticipating the later attachment of intent-driven prostheses, they will strive

An icon for the women's restroom at Össur.

to preserve muscle activity at the end of amputated limbs in ways that may enable the use of these "smart" prosthetic limbs. As I'll shortly explain, this is exactly what Hugh Herr's current project has taken on.

By restoring close to full mobility and functionality to amputees, and by creating a design that can reach a broad market, Össur has changed the way we think about prostheses. Instead of signaling disability, they signal normality. That's the message I got when I saw a sign in Össur's offices directing me to the women's room. It showed the standard skirted figure representing a woman, but the figure sported a prosthetic leg.

o

My visits with Herr and Össur gave me ample reasons to believe that we are well along on the path to solving the hard problems for replacement limbs. But many disabilities arise from damage not to limbs but to the nervous system itself. Restoring mobility after brain injury is a wildly complex problem, but here, too, we're making impressive progress. I learned just how impressive when I traveled to Geneva, Switzerland, to catch up with John Donoghue, one of the world's leading systems neurophysiologists.

Donoghue, a friend and colleague from early in my neuroscience career, has devoted his career to understanding the cerebral cortex, the part of our brains that most distinguishes us from other animals. His work has focused on the motor cortex, the set of nerve cells in the cerebral cortex and their connections, which directs our movements. These days he divides his time between Brown University, in Providence, Rhode Island, where he is a professor, and Geneva, where since 2014 he has held the role of the founding director of the Wyss Center for Bio and Neuroengineering. At the Center, Donoghue leads a team of engineers, biologists, computer scientists, and clinicians in a collaborative effort to design devices that will return mobility to people paralyzed by brain injury or disease.

Research over the last century has provided a basic understanding of how the brain orchestrates physical movements. After we've decided, consciously or unconsciously, to make a movement—for example, raising a hand to answer a question, or taking the first step to descend the stairs for breakfast—a part of the brain called the primary motor cortex (PMC) goes to work. The PMC sits on the surface of the brain, just above the front of our ears. Nerve cells in the PMC have very long axons that function like long wires to carry signals from the PMC down through the base of the brain and through the spinal cord to reach motor neurons at the right level of the spinal cord to carry out a specific task. When the motor neurons receive input from the PMC neurons, their axons relay that signal out of the spinal cord to the muscles that carry out the intended movement. To raise my hand, nerve cells in the PMC send their signal to motor neurons in the arm region of my spinal cord, and the axons of the motor neurons, in turn, carry the signals out of the spinal cord to the muscles that control the movement of my arm.

When I was studying neurobiology in graduate school, my textbooks described the PMC as having a "pointillist" organization: PMC nerve cells, it was believed, were arranged in a kind of map in the brain that mirrored the body's surface, and neurons at each point on the map would drive the motor neurons for a specific muscle in the body. Early in his career, however, Donoghue did research that changed the way we understand the organization of the PMC and how the brain drives physical actions. In a series of groundbreaking studies, he showed that each point in the PMC corresponds not to a particular muscle or set of muscles but to a complete action. When I raise my hand, for example, one location in the PMC drives the muscles in both my back and in my arm, coordinating their outputs to make a smooth motor performance.

Donoghue achieved worldwide recognition for his studies. If he had followed the usual path, he would have continued his research on the basic neurobiology of the cortex. But instead he decided to pursue an almost inconceivably bold ambition: to use his understanding of the PMC and its organization to help to return the power of movement to people paralyzed by spinal cord injury and disease.

When the brain and spinal cord are intact, neurons (nerve cells) in the PMC and other parts of the brain send signals through

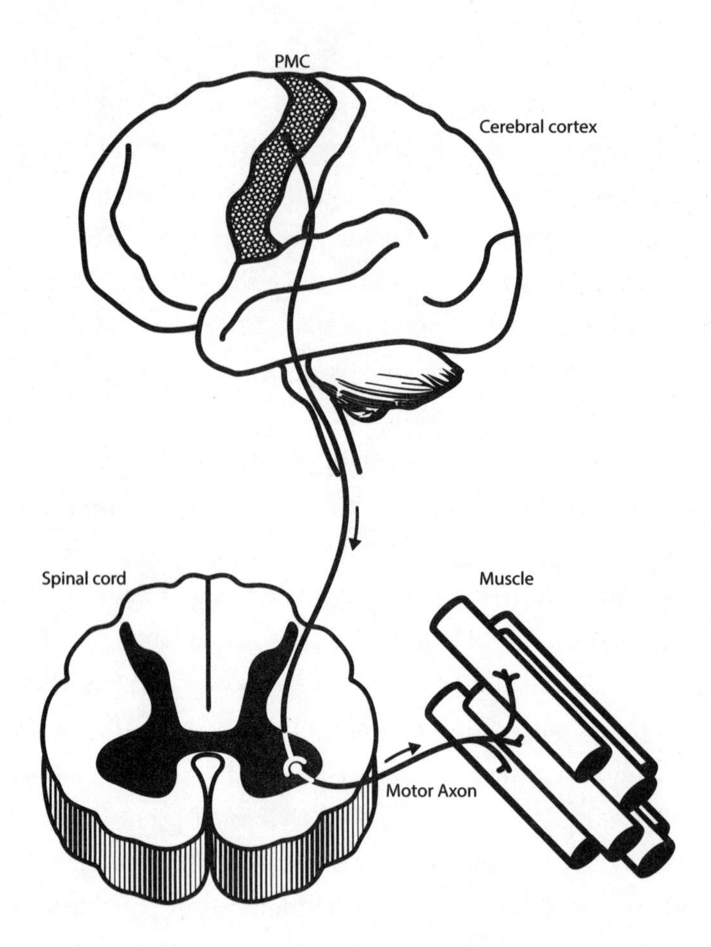

Activity in the primary motor cortex (PMC) drives motor movements. Neurons (nerve cells) in the PMC have long axons that extend from the PMC into the spinal cord. When the PMC neurons are active, their axon endings in the spinal cord activate motor neurons. The axons of the motor neurons connect to muscles and drive muscle movements.

their axons that drive and coordinate movement. These signals travel along axons from the brain down through the spinal cord to activate motor neurons. Recall that motor neurons in the spinal cord have axons that extend out of the spinal cord to activate muscles. In order for our movements to be effective and coordinated, they must be perceived, setting up a kind of feedback loop. We perceive our surroundings in the sensations of touch and of the position of our arms and legs through signals that travel in the opposite direction, into the spinal cord, along axons that connect sense organs in the muscles, skin, and joints to nerve cells in the spinal cord and then up through the spinal cord to the brain. The sensory connections that travel back to the brain give us an awareness of pain, heat, or cold or of the position of an arm or leg.

If we suffer an injury to the spinal cord, however, all of these connections can be cut, disconnecting the brain from muscles and sensing organs, resulting in paralysis and numbness. The spinal cord is protected from injury by its bony covering, the bumpy

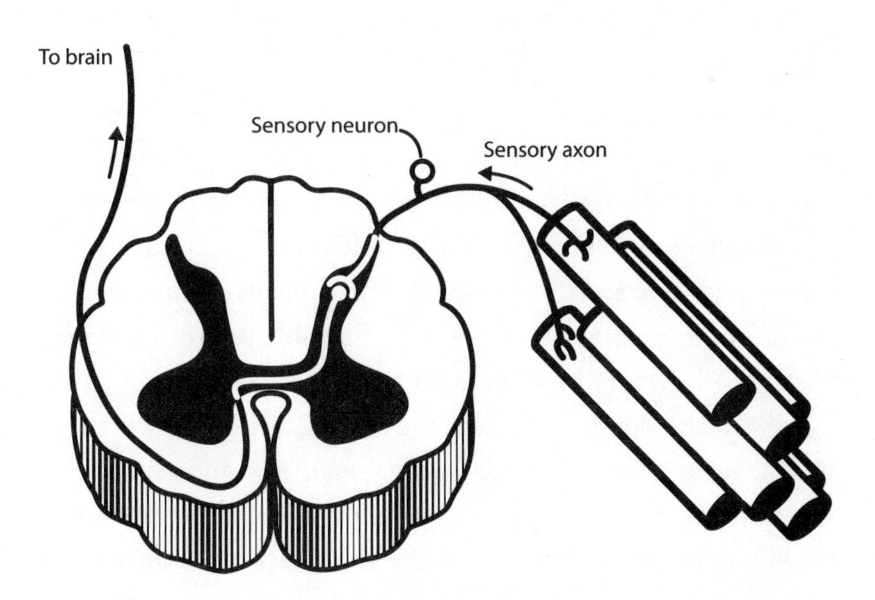

Sensory information from the muscles (or skin) travels into the spinal cord through the axons of sensory neurons, which sit outside the spinal cord. The incoming signals activate spinal cord neurons that carry the sensory signals to the brain.

vertebral column that we see and feel along the center of the back. The brain is, similarly, protected by the skull, its own hard bone covering. Protecting the brain and spinal cord is essential: unlike skin, bones, liver, and muscles, the nerve cells in the central nervous system do not effectively recover from injury. All our abilities and actions—for example, breathing, talking, seeing, walking—are driven by neurons that if damaged cannot repair themselves. Despite an enormous amount of study and research, we do not yet understand sufficiently the biology that prevents the brain from recovering, and so we do not yet have effective strategies for healing brain injuries. That means that severe spinal cord injuries often make intentionally moving a leg or hand forever impossible. With training, some individuals can recover some function through pathways that were not damaged by the injury, but many people with spinal cord injuries remain paralyzed and without sensation.

Donoghue recognized the possibility that, despite the disconnection that prevents voluntary movement following a spinal cord injury, the PMC neurons might still be active. He wondered if he could record the activity of the PMC neurons to understand someone's intent and then devise alternate means to translate that intent into movement. If so, he might be able to provide people paralyzed by injury or disease with a way of reengaging physically with the world. To explore the possibilities, he joined forces with an array of colleagues, physicians, engineers, and neurobiologists. Together, in the early 2000s, they developed an intracortical brain-computer interface (iBCI), which records brain activity and relays it to a computer that in turn uses those recordings to drive movement.

Donoghue and his colleagues first showcased the potential of their iBCI in 2006, when a young man, paralyzed from the neck

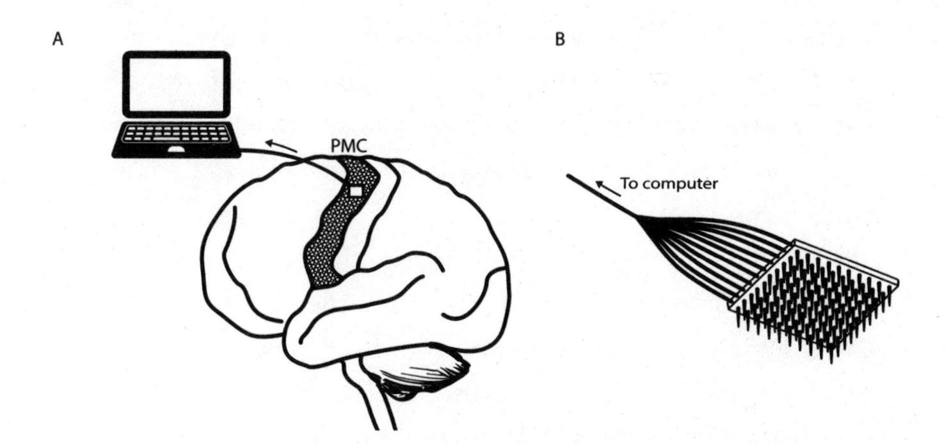

An intracortical brain-computer interface (iBCI) can record intended movements from neurons in the primary motor cortex (PMC). (A) A small electrode array (white box on the PMC) records the activity of PMC neurons. When the intent to move a limb is initiated, the electrode array detects signals in PMC neurons; those signals are transmitted to a computer. The computer decodes the signals and relays them to an external device, like a robotic arm. (B) The iBCI electrode array has roughly a hundred very thin, needle-like electrodes that are implanted into the PMC. The electrodes record the activity of PMC neurons and transmit those signals through small, flexible wires to a nearby computer.

down by a spinal-cord injury, used an iBCI with a computer interface to play the computer game Pong. As he imagined moving a computer mouse with his hand, the iBCI read his intentions and translated them into moving the Pong paddle on a computer monitor.

Donoghue's team made remarkable advances in the years that followed, and in 2012 they reported that a woman named Cathy, paralyzed from the neck down by a stroke, was able with the iBCI to direct the movement of a robotic arm simply by thinking about moving her own arm. What made this possible was a very small chip about the size of a baby aspirin that the team had planted in the area of Cathy's PMC that would have initiated her natural arm's movements. The chip held close to a hundred tiny, needle-like electrodes that inserted into Cathy's PMC. Each electrode recorded electrical signals coming from the PMC neurons closest to it and then sent those signals through a very slender wire to a computer, which in turn used the signals to decode Cathy's intention. That movement command code was then sent to a computer-driven robotic arm. For the first time in the fifteen years since her stroke, Cathy could move objects, simply by thinking about it.

The team captured Cathy on video successfully using a robotic arm. Her task was to pick up a water bottle containing her favorite morning drink, a cinnamon latte, and drink from it. In the video, she furrows her brow in concentration, as she wills the arm to move. It works: the robotic arm slowly grasps the bottle, picks it up, and brings it to her mouth. The bottle contains a straw, and as she sips from it and tastes the latte, a slight smile appears on one side of her mouth. After a short while, she wills the arm to pull the bottle away and place it upright on the table in front of her, at which point she turns her head toward the camera with a triumphant grin. It's a moment that brings tears to my eyes.

In 2017, Donoghue and his team took the technology a major step further, by enabling Bill, who had been paralyzed for eight years after a cycling accident, to move his own arm. The team implanted two iBCIs in Bill's PMC and a set of stimulating electrodes in his paralyzed arm muscles. Bill, like Cathy, was asked to grasp a cup, bring it to his mouth, and drink from it. In this case, when Bill willed his arm to move, the signals the iBCIs picked up from activated PMC neurons were relayed to a computer, which sent its message to the electrodes in Bill's arm muscles, directing them to behave as they would normally. And, again, it worked: Bill willed his hand to grab the handle of the cup and brought it to his mouth for a sip, the first time in more than eight years that he had been able to use his arm to take a drink or feed himself.

When I visited Donoghue at the Wyss Center in the fall of 2017, he delightedly showed me what he calls his "museum": a glass case displaying an array of bionic devices. Many of the devices were the predecessors to his iBCIs: for example, cochlear implants that now allow many deaf people to hear and atrial defibrillators that normalize abnormal heart rhythms. He showed me how much these life-transforming technologies had become smaller over time, thanks to biological insights and technological developments. And he handed me the newest version of the chip that sits in Bill's PMC.

The chip measures about 4 millimeters (mm) on a side. It has a hundred hairlike electrodes attached to it, each 1.5 mm long, which give it the look of a tiny hairbrush with fine, barely visible bristles. At the base of the chip hangs a slender, very flexible wire, which allows the electrodes to send the signals they receive to the computer that decodes their signals into electronic output to drive intentional movements of a robotic or natural arm. When I took the device in my hand, I was struck by how little it weighed and by how much it nevertheless can make possible.

Next-generation iBCIs and their chips will be much smaller and entirely wireless. Some engineers anticipate sensors smaller than a grain of table salt that will sense and transmit neuronal activity. Devices that small could be placed throughout the brain, recording a much larger array of signals and providing a more accurate reading of the individual's intent.

Since the late 1960s, when brain-computer interfaces were a new and startling idea, the rapid development of both neurobiological insights and computational power promises new technologies that, in the not so distant future, may be able to alleviate the most daunting disorders. One of Donoghue's colleagues, Dr. Leigh Hochberg, envisions that the next generation of wireless devices will record brain activity and have the capacity to decipher the onset of abnormal patterns in, say, epilepsy or manic-depressive disorders. He anticipates devices that would send corrective signals back to the brain to restore normal patterns of activity—and, in doing so, restore normal life to those disabled by neurologic and psychiatric disorders.

o

Medical miracles like these are nearer than we might think. Hugh Herr and his team have begun first-in-human projects to restore more natural mobility to amputees. The team brings together orthopedic surgeons, neurobiologists, mechanical and electrical engineers, and molecular geneticists, along with student colleagues whose professional directions will mix all these fields in new ways. Like the groups at the Wyss Center and Össur, they are using normal biology to recover motor control. They were ready to move forward when Jim Ewing contacted Hugh Herr. Ewing volunteered to be the first person to receive, what has now been named, the "Ewing Amputation."

Herr and his colleagues have focused their efforts on re-creating the normal "agonist/antagonist" pairing of muscles around joints and then directing normal nerves to connect these muscle pairs to the spinal cord. When you bend or extend your ankle, for example, muscles on the front and back of your calf take turns contracting or relaxing; both contraction and relaxation are controlled through neural circuits that travel from the muscle to the spinal cord, through a series of relays within the spinal cord and then back to the muscle. Coordinated contraction and relaxation are required for smooth and effective movement. The sensory input from the muscles and joints into the spinal cord engages local circuits to coordinate movements and also travels to the brain to provide awareness of the position of the leg and foot.

To prepare amputees for their new devices, Herr and his surgical colleagues had to redesign the amputation procedure. In it, they reposition the remaining lower leg muscles that had controlled ankle movements, along with the attached nerves. They create agonist/antagonist pairs by using a tendon to connect pairs of muscles that would have extended and flexed the ankle. Following this surgical procedure, when amputees think about flexing or extending their ankle, their leg muscles contract and relax in the normal patterns as agonist/antagonist pairs. Electrodes over the muscles detect their activity and transmit those signals to a computerized ankle that responds to the signals, reproducing the action of an intact ankle. Like a normal leg, the wearer's intent drives muscle movements, but rather than directly driving an action, the ankle's computers translate the muscle signals into actions of the prosthetic ankle and foot.

By 2015, Herr's team had moved through all the preparation for this potentially life-transforming procedure, including device design, computer modeling, and experimental testing; they were ready to move into a human trial when Ewing contacted Herr. Ewing faced a very difficult decision: Should he give up his natural leg, which after a full year still generated intolerable pain and couldn't support his activities of daily living, not to mention his athletic ambitions? His physicians gave him a choice: continue to try to resurrect his ankle or have his lower leg amputated and turn to prosthetic devices. Ewing describes the difficulty of the decision: "Amputation felt drastic, but the alternative prospects were dismal." Herr recounted his own experience with prosthetic legs, as well as the evolution of the devices in his and others' labs. Over the course of numerous conversations, consultations, and demonstrations, Ewing decided on amputation. He volunteered to be the first human to undergo the new procedure that would enable him to use Herr's newest device—namely, a prosthetic lower leg, ankle, and foot that he could both move and feel.

On a normal day, Ewing wears currently available prosthetic legs to walk, run, climb, ski, or scuba dive. He has resumed his previous active life without pain. On an extra-normal day, he joins the Herr team to pioneer his new brain-animated leg. On my recent visit to Herr's Biomechatronics lab, I watched a video of Ewing climbing a rock face in the Cayman Islands. Like all climbers, his gaze looks up as his left foot searches for its toehold; without any visual guidance his prosthetic toe finds a spot, and then his new leg supports his body's weight and balance as his natural, right leg and foot find the next toehold. Reaching the top of the cliff, Ewing sits and looks out at the ocean scene, as he raises his leg to rest his left foot on a convenient outcropping.

With this complex bio-computational-mechanical surgery and device, Ewing has recovered a remarkable array of nearly natural movements. He reports that he "feels everything, just as though the prosthesis was a part of me." Herr anticipates that within twenty years, getting a prosthetic "will be nearly like getting your biological limb back."

In the not-too-distant future people who have suffered disabling injuries will walk, talk, and engage with the world again. To give MIT's governing board a glimpse of that future, in 2010 I invited Hugh Herr to speak to them about his new technologies.

We met in a small campus auditorium. After I introduced him, Herr walked to the front of the auditorium with the graceful saunter of an athlete. Nobody had any reason to suspect that he was a double amputee.

As Herr began to speak, he looked out over the audience and observed that many of the people in the room were wearing a transformational prosthetic technology: eyeglasses. Poor vision can be debilitating, he noted, but we don't consider people with bad eyesight disabled. Why? "Because we have terrific technology that enables people with poor eyesight to lead a perfectly functional life," he said. The analogy captured perfectly what Herr hopes to accomplish with his prostheses. As he began talking about his work, Herr rolled up first one trouser leg and then the other, gradually revealing, to the audience's great surprise, his prosthetic legs, ankles, and feet. I doubt anybody who had watched Herr walk into the room and start talking considered him disabled, for the simple reason that he was wearing prostheses that did their job as well as eyeglasses do theirs. Standing on state-of-the-art prosthetic legs and computer-assisted ankles, Herr concluded his talk with an exhortation. "We call a condition a disability only so long as the available technology is insufficient to overcome it," he said. "Let's take amputation and paralysis off the disabilities list."

6

FEEDING THE WORLD

In the fall of 2017, I found myself in a darkened vestibule at the Danforth Plant Science Center, just outside of downtown St. Louis, peering through a small window into a 750-square-foot "growth house." What I saw inside looked almost like a cartoon: under bright lights that made their leaves glow a supernatural shade of green, about a thousand small and medium-sized plants were doing a carefully choreographed dance.

In reality, they were moving along 180 meters of conveyor belts. Emerging from a central array, plant after plant moved almost jauntily, switching to a new belt here, merging with another there, and stopping periodically at different stations as they made their way around the room. Each station had a special function. One delivered a customized allotment of water to each plant. Another did the same for fertilizer. One recorded weight. Another took digital photographs from a variety of angles, recording such features as height, girth, and leaf number, along with their branching pattern and size. One took near-infrared photographs to record water content. On and on the dance went, until the plants returned to their places in the central array, ready to begin the process again the next day. The lighting, temperature, and humidity in the room were set precisely and controlled carefully, giving the scientists in charge of the room the ability to push temperatures to extremes (for stress testing) and to manipulate the light spectrum and light intensity (for the testing of photosynthetic capability, shade adaptation, and more).

I watched for a few minutes, absorbed by the strangeness of the sight. But this wasn't just a show put on for my benefit. These plants were doing a dance that would go on for weeks, the whole process carefully developed by a part of the Danforth Center known as the Bellwether Foundation Phenotyping Facility. Each plant was carefully monitored, carrying a radio-frequency identification (RFID) chip and bar code monitor so that the facility, without any human intervention, could ensure that all of those images, detailed measurements, and customized allotments of fertilizer and water were accurately recorded and attributed—often for the entire life of the plant. This work brings biology and engineering together to help us understand in powerful new ways how a plant's genetic endowment interacts with its environment to produce all of its traits and properties over time, together known as its phenotype.

The effort is ambitious: Danforth scientists collect both the genetic and the phenotypic data for their plants in forms amenable to large-scale computational analysis. In essence, this means translating all of the plants' relevant genetic and phenotypic properties into numbers, and from those numbers deriving a map of how the plants' traits are constructed. Just as genomics refers to the full set of genetic information of an organism, phenomics refers to the full set of phenotypic information, its physical traits.

The amount of data that the growth house captures for every

plant that passes through it is mind-boggling. Tens of thousands of genes orchestrate the development and function of a plant, with the many possible combinations of subtle variations among genes creating an almost endless array of possible phenotypes. Never before have scientists and engineers been able to gather anything like this quality and quantity of information about plants and how they grow. Thanks to a new generation of plant scientists at the Danforth Center and elsewhere, we're developing a better picture than ever before of the complex ways in which plants express their genes over time. And due to the advent of big data and high-powered computer analytics, we're fast learning how to study, manipulate, and record phenotypes in ways that will allow us to engineer plant variants that will dramatically improve crop yield.

This convergence of biology and engineering is different from the examples we've looked at in previous chapters. In those cases, we saw how scientists and engineers are harnessing biological organisms and mechanisms to help us solve a variety of technical challenges. In this case, however, we'll see how data gathering and computational engineering tools can lend insight into complex biological characteristics. For the development of new plant varieties and new crops, these tools are providing a new level of understanding of the physical properties of plants and how those properties develop and change over time. This kind of information promises the ability to select plants with optimal traits far more efficiently and accurately than we do now. This convergence is every bit as revolutionary as the others we've looked at: it has set in motion new approaches to agriculture and food production that will help us meet the urgent demand of feeding our increasing, and increasingly prosperous, global human population that, by 2050, is projected to reach 9.5 billion or more.

o

The demand for food will be vast. To meet it, we will have to almost double our present global crop productivity. Will that require a doubling of the amount of land we farm? Or could we instead increase the productivity of our current farmland through technological innovation? This challenge is a daunting one—but it's one we've overcome before. During the past hundred years, in fact, we've doubled how much corn we produce not just once but four times, as a result of better field management through crop rotation, the increased availability of natural and synthetic fertilizers, the improvement of crop genetics through selection and genetic engineering, and increasingly efficient agricultural mechanization.

We've been engineering increases in food productivity for a very long time. Evidence from archeological excavations indicates that humans began engineering food crops more than ten thousand years ago in the Fertile Crescent. Our early ancestors probably gathered and planted seeds from plants that grew in the wild; then they improved their productivity by carefully observing, selecting, and propagating the most propitious variants. They were early genetic engineers despite knowing nothing about genes.

Genetics is a modern science. The word *gene* dates back only to 1905 when the Danish botanist Wilhelm Johannsen used it to refer to a discrete inherited unit that participates in determining observable physical traits. Johannsen derived the word from *pangene,* a term coined twenty years earlier by another Danish botanist, Hugo de Vries. But the existence of such an inherited unit had in fact first been proposed even earlier, by the Augustinian friar Gregor Mendel, the man generally considered to be the founder of modern genetics.

Between 1856 and 1863, working in obscurity at the Augustinian Abbey of St. Thomas in Brno (now part of the Czech Republic), Mendel crossbred thousands of pea plants and discovered that their traits sorted from one generation to the next according to a set of predictable mathematical rules. These rules, Mendel suggested, could be explained by the existence of discrete inherited units, which he called "factors," that governed the inheritance of a plant's physical characteristics. The rules that Mendel discerned, today known as Mendel's laws of inheritance, laid the foundation for modern genetics.

Mendel published his remarkable deductions in 1865. Key among them was that a given physical trait for a plant—seed color, for example—is determined by the expression of two genes (as we now call them), with one of the pair coming from each of the parents. The two genes of a pair can contribute differently to the expression of a trait, with one being dominant and the other recessive. In the case of the genes for the seed color of pea plants, yellow is dominant over green. If a pea plant inherits two genes for yellow colored seeds, one from each parent, it will have yellow seeds; if it inherits two genes for green colored seeds, it will have green seeds; but if it inherits one gene for yellow and one for green seeds, it will have yellow seeds, because yellow is dominant.

When Mendel described the "factors" that lay behind his laws of inheritance, he had no understanding of what physical form they might take. The same was true for Johannsen: when he first used the term *gene*, like Mendel, he was describing a behavior, not a physical structure. The actual structure of DNA and its role in determining an organism's traits would not be discovered until almost fifty years later.

This pattern of scientific progress is typical, and it's a pattern we've encountered throughout this book. It begins with the keen

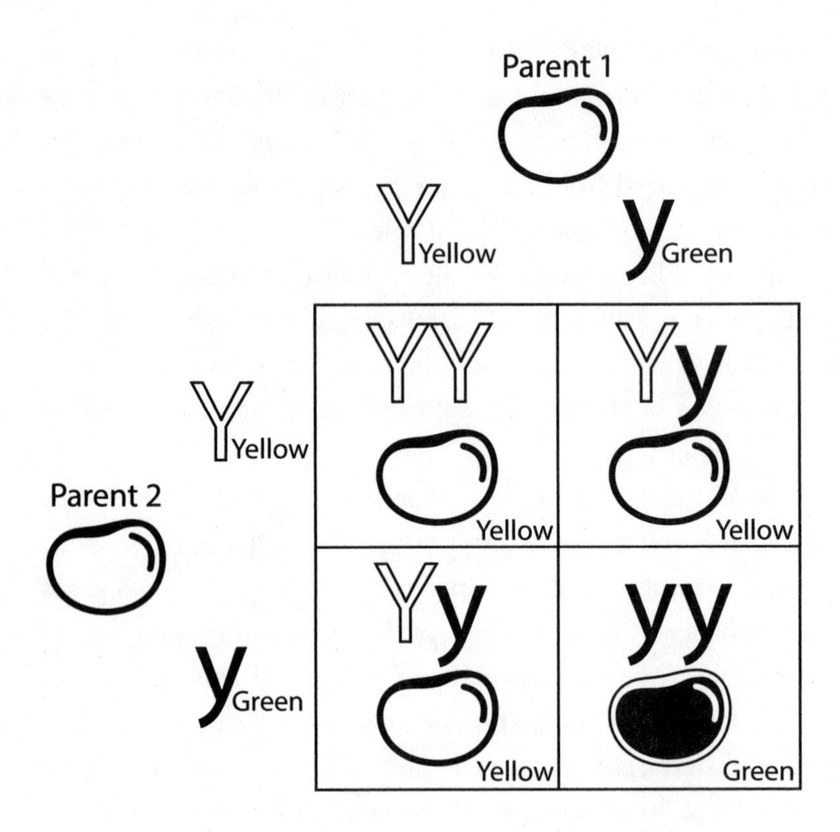

The Punnett square is the standard representation of a genetic cross. In pea plants, yellow seed color (Y) is dominant over green (y). Pea plants with either two yellow genes (YY) or with one yellow and one green gene (Yy, heterozygotes) will have yellow seeds. When two parent Yy pea plants are crossed to each other, the offspring plants that inherit either one or two dominant, yellow (Y) genes will have yellow seeds. That is, offspring that inherit two Y genes (YY, homozygotes) will have yellow seeds, as will offspring plants that inherit one yellow (Y) and one green (y) gene (Yy, heterozygotes). Only the offspring plants that inherit two copies of the recessive green gene (yy, homozygotes) will have green seeds.

observation of a phenomenon, which through subsequent exploration leads to the discovery of physical components that give rise to the phenomenon. Michael Faraday described electromagnetic forces in the mid-nineteenth century, long before J. J. Thomson's 1897 discovery of the electron as the particle that gives rise to those forces. Similarly, the kinetics of how water crosses a cell membrane were described in great detail in the first half of the twentieth century, long before Peter Agre discovered the water-channel protein that provides the physical conduit for water molecules to cross membranes.

Mendel's work was poorly understood and mostly forgotten during his lifetime, but at the beginning of the twentieth century, several researchers rediscovered and replicated it and, following Johannsen, started calling Mendel's "factors" *genes*. Guided by Mendel's laws and the emerging field of genetics, agronomists began to think in new ways about how to breed plants.

Meanwhile, other researchers were gradually ramping up the hunt for the molecular structure of DNA—a hunt that had begun in 1869, when the Swiss biologist Friedrich Miescher had isolated a substance that he called "nuclein" from the cell nucleus of blood cells. Miescher didn't know it, but he had identified the physical substrate of heredity. In fact, for decades most scientists believed that nuclein couldn't possibly carry genetic information because its structure was too repetitive and boring. The rich diversity of observed phenotypes, they reasoned, should require an equally diverse material, and so nuclein's relevance to heredity was largely dismissed.

The hunt for the substance of heredity continued for decades, until 1953, when in one of the most famous papers ever published, James Watson and Francis Crick proposed the double helix as a model for the molecular structure of DNA. As we saw in chapter

2, the double helix that they described resembled a twisting ladder with parallel phosphate-sugar chains interconnected by rungs composed of pairs of the four bases that always pair in the same pattern (guanine with cytosine and adenine with thymine), thereby mediating DNA's accurate replication.

With the description of the structure of DNA, the field of molecular biology truly came into its own. Now that biologists understood its basic structure, they were soon able to determine that genetic information is first copied from DNA into RNA (very similar in structure to a single strand of DNA) and then translated from RNA into proteins. The proteins then guide the assembly of cells into tissues and give rise to many of the emergent properties of tissues and whole organisms.

It was a landmark moment in the history of science and agriculture. After Watson and Crick's discovery, molecular biologists deployed their new understanding of genetics to imagine entirely new approaches to growing food. By 1983, scientists established methods for transferring DNA into plant cells as well as genetically engineering insect-resistant tobacco. All sorts of advances now became possible. Adding genes to a plant might enhance its nutritional content, reduce its need for water, or decrease its susceptibility to disease.

In the decades that followed, plant geneticists and agronomists have developed high-yield strains of corn, wheat, rice, and other crops. These innovations—along with increasingly efficient fertilizers, irrigation systems, crop-rotation techniques, and mechanized farm equipment—have led to great surges in crop productivity. From the 1860s to the late 1930s, for example, corn production in the United States hovered at around 30 bushels per acre, whereas today that figure consistently reaches above 150 bushels per acre. Such leaps made it possible in the latter half of the twentieth century to cheaply, safely, and reliably feed millions of previously undernourished people and animals around the world.

Despite the stunning increases in agricultural output that these innovations have made possible, roughly 800 million people still live in food scarcity globally, and over 3 million children under the age of five die each year of starvation. Plant scientists continue to explore how the manipulation of genes can improve crop productivity, and their efforts are producing some remarkable results. Combining those remarkable results with the power of nature's genius expressed in natural plant variations promises even greater increases in crop productivity. Norman Borlaug, one of the most important architects of the green revolution, noted in 2002 that if we aim to accelerate crop improvements to meet our anticipated population growth, we will need "both conventional breeding and biotechnology methodologies."

o

In 1994, the FDA approved the first genetically modified food plant for sale: the FlavrSavr tomato. Plant biologists engineered the FlavrSavr to ripen more slowly than conventional tomatoes by adding a gene that inhibits the carefully regulated production of an enzyme in tomatoes that breaks down pectin. (Recall that enzymes are proteins that cut other molecules.) With that gene in place, the FlavrSavr variant could be left to mature longer on the vine and would arrive on a grocer's shelf with better flavor and less damage. The FlavrSavr didn't become a commercial success, but it did demonstrate that direct manipulation of plant genes could improve food crops and that the Food and Drug Administration (FDA) could evaluate such foods for health risks.

Today, the majority of corn grown in the United States, along with the majority of cotton and soybeans, carries genetic modifications that make it resistant to insect pests and herbicides. As recently reported by the National Academy of Sciences (NAS), these modifications have led to many benefits, among them a reduction in the amount of pesticides and herbicides used in growing these modified crops and in the cultivation of nonmodified crops grown in their vicinity. The NAS report also did not find evidence for adverse health effects from these crops when they were grown using current best practices.

Molecular biology and genomics have allowed us to identify specific genes in plants that code for specific proteins—proteins that can, for example, protect plants from insect or viral pests, or provide resistance to drought or frost. Similarly, we have identified specific gene mutations as the cause of some human diseases, allowing us to design drugs that can provide effective therapy and even cures in a small number of cases. But most traits and most diseases are not products of single genes or single proteins. Instead, they have genetic underpinnings that are far more complex. Analyzing those complexities has required another kind of conversation between biology and engineering, one that engages state-of-the-art biological technology with state-of-the-art computation.

Consider just one example: the sequencing of the human genome. This job, considered almost inconceivably difficult just thirty years ago, was made possible by incredibly rapid advances in the technology for determining the sequences of bases in DNA and by equally rapid advances in computation. Using as a baseline the information from the Human Genome Project, work going on around the world today seeks to identify genetic factors for diseases and resistance to disease. This kind of detective work often requires comparisons among tens of thousands of individual genomes. A project to discover the genes that could predispose an individual to autism or schizophrenia, for example, might survey the genomes of more than 60,000 people and calls on advanced bioinformatics—computational engineering—developed in concert with biologists.

The same type of detective work goes into discovering the key genes for beneficial plant characteristics. By manipulating single genes, biologists and engineers have made impressive progress in producing high-yield crops with a lower need for water, fertilizer, and insecticides. Corn and other crops, for example, have been modified to include a gene that produces a natural insecticide, a protein endotoxin produced by a soil-dwelling bacterium, *Bacillus thuringiensis* (Bt). Organic farmers commonly use preparations of Bt proteins; sprayed on crops, they act as an insecticide to prevent damage from some particularly crop-damaging insects but without any demonstrated ill effects on other insects, humans, or other animals. Corn and other crops have been engineered to express a gene for a Bt endotoxin, making the crops resistant to attack by corn borers, rootworm larvae, and other insects. Bt variants of cotton and soybeans are also widely cultivated and have reduced the use of insecticides. Recent analyses have found a further benefit, in that nearby, non-Bt-carrying crops require less insecticide, perhaps because the Bt-carrying crops have reduced the local insect population.

Another widely deployed genetic modification makes a plant resistant to a particularly effective herbicide, glyphosate. Also known by its trade name, Roundup, glyphosate effectively blocks the production of amino acids in plants (but not in insects, birds, mammals, or other animals) essential for protein production. This causes rapid death of the plants that come into contact with the herbicide. A field of herbicide-resistant corn (HT corn), for example, can be treated with glyphosate to control the growth of weeds without damaging the corn. This kind of weed control reduces the requirement for tilling, which is important to reduce the loss of topsoil. The widespread cultivation of HT crops has reduced the use of herbicides. And, while concerns about glyphosate's safety have continued, a recent review by the National Academy of Sciences found that adverse effects have not accompanied appropriate use.

Another advance in agriculture has included new strategies to improve the nutritional value of plants. By adding two genes to the rice genome that increased the synthesis of vitamin A, plant scientists have created a variant known as Golden Rice, significantly enhancing rice's nutritional value. This innovation has the potential to save many lives in developing countries where insufficiency of vitamin A contributes to the deaths of more than 500,000 children a year. Despite numerous studies demonstrating the safety and benefits of Golden Rice, its use has been inhibited by unsubstantiated concerns about genetically modified food. These concerns have been raised in countries where the rice might be grown and—perhaps even more volubly—in countries in the developed world. After much scientific study and debate, Australia and New Zealand approved Golden Rice in 2018, but countries that could most benefit, like Bangladesh and the Philippines, have yet to approve this lifesaving staple.

These kinds of concerns about the safety and economics of genetically modified crops have slowed their deployment. Up to a point, this is reasonable. Nobody, after all, wants to create crops that turn out to have inadvertently and irreversibly harmful effects. But the 2016 study by the National Academy of Sciences suggests that we're stalling and regulating to a degree that itself is causing harm. The study found that genetically engineered crops are safe when grown responsibly according to a set of guidelines that the NAS laid out in its study, which means that many potentially lifesaving crops could today be deployed in the developing world. This includes crops such as cassava, which have too little economic return to support the needed investment to complete complex regulatory approvals.

o

Despite the many gene-by-gene advances we have made in recent decades, we still have a lot of trouble identifying key genes and gene interactions that control the complex traits that affect agricultural productivity, like drought resistance or increased grain size. Fortunately, we're increasingly able to turn to a different method of analysis through the use of new technologies to screen the phenotypes of large numbers of plant variants. This is the approach used at the Danforth Center and facilities like it, where state-of-the-art imaging and computing technologies are used to record, analyze, and compare the physical characteristics of hundreds and even thousands of plants. This process, known as high-throughput phenotyping, has been developed for lab-based environments and is now under development for field experiments, the critical test case for agricultural crops.

The process of genetically engineering or conventionally breeding plants to express desired properties remains an extremely challenging proposition. As helpful and as productive as current approaches are for genetically modifying plants, the difficulty in identifying key genes for complex traits has limited the possibilities for generating crops with intrinsically increased food productivity. So to increase agricultural productivity further and faster, we've recently turned to a different method of analysis: using modern engineering technologies to screen hundreds or thousands of plant variants to identify the few that possess the desired traits.

In effect, we are returning to earlier agricultural methods carried out by farmers for selecting and propagating more productive plant variants. But we now need methods that can sort through large populations of candidate plants to monitor and sort them based not just on whether they express individual genes but also whether they express the full set of desired physical characteristics over time—their phenotypes, that is.

The challenge increases exponentially as breeders incorporate wild relatives with promising traits, including resistance to nematodes (a crop-destroying organism). Wild soybeans from China can contribute nematode resistance when crossed with a productive variant widely grown in the United States, for example. The combination of the two genotypes, mixing close to 40,000 different genes, creates a mind-boggling array of phenotypes, only a few of which have optimal traits. Automated phenotyping accompanied by massive computational processing has the potential to identify the proverbial needle in a haystack. But even that provides only a starting point. Generations of breeding and selection will have to follow before a farmer plants fields using seed that yields a uniform, reliable crop. This is where convergence technologies help: they can substantively accelerate the time from breeding to harvesting a new, more resilient, and more productive variety of a crop.

This new approach uses state-of-the-art technology: first, to record physical characteristics with image-based plant phenotyping methods; and second, to expand these imaging technologies to the scale required to select among hundreds or thousands of plants for high-throughput phenotyping. High-throughput methods have been developed for lab-based environments, like the Bellwether Foundation Phenotyping Facility at the Danforth Center, and are under development for field experiments. Because phenotypic traits manifest themselves over the full life cycle of a plant, these engineering technologies must also have the capability to repeatedly measure the traits for a single plant and accurately assemble its full phenotype, which can then be compared with the phenotypes of other plants.

Unlike genomics, the data collected in phenomics are nonlinear and evolve in a variety of ways across both time and space. What this means is that in the search for sweeter corn or drought-resistant wheat, we must monitor the entire life cycle of plants and study all sorts of questions, fully evaluating the phenotypes of these plants before we can confidently select the most viable and productive variants. How do the plants behave in response to too much or too little rain? Will they thrive with more or less fertilizer? How much of their food products can be harvested? What is their nutritional value? What do they look and taste like?

Until recently, we've been constrained in how we can gather and study this sort of information. Barbara McClintock, the great twentieth-century corn geneticist, famously did it on foot and by hand, walking between rows of the corn plants that she had hybridized, feeling the texture of their leaves, and noting the size and color of the developing ears, and then carefully noting their seed patterns.

Today, scientists and farmers are turning to drones and satellite imagery to do some of this work: monitoring their fields regularly over hundreds of acres of crops; using them to hover over and document the conditions in particular areas of interest or concern; and even equipping them with highly specialized cameras so that they can measure how much light plants are using for photosynthesis or how much water they need. But even this kind of data gathering has its limits. Monitoring single plants in the field is still pushing the limits of current technology, and following an individual plant over the full course of its development is particularly challenging. High-throughput phenotyping with a focus on individual plants has, in fact, become the rate-limiting step in using large-scale genomic data to improve crops.

To see what I mean, imagine conducting an experiment to test whether crossing two distantly related plant variants will yield a new variant that can thrive with less water. This experiment is a bit like Gregor Mendel's experiments in crossing pea plants but with an important difference. Mendel used highly inbred plants, which limited the range of possible outcomes in their offspring. Even with that limited set of possible outcomes, he had to examine hundreds of plants.

In our experiment, the parent plants would contribute a wider range of genetic information, which should produce a larger range of phenotypes among the offspring—some, we hope, that are more tolerant to drought. The challenge is that at the outset we don't know which genes, or how many of them, contribute to drought tolerance. Let's say that our best guess (from previous analysis of the parent plants) is that 1 percent of the offspring might display the trait we seek. We might want to test the effect of water deprivation at different times during the plant's life. And if we're interested in a food product—a tomato, say, or a soybean—we'll want to follow the plants from germination through to the production of that product. A rare effect that appears in, say, 1 percent of progeny would probably require a survey of hundreds of plants several times over the course of the growth period. To include variables like water, light, and temperature sensitivity, thousands of plants would need to be monitored. What might seem to be a simple experiment rapidly becomes very complex.

Until recently, we could only do this in the manner of Barbara McClintock—that is, in person, by visually monitoring the progress of plants one by one, day by day. But today we're developing observational techniques and technologies that allow us to monitor and sort plants much more quickly and efficiently based on their phenotypes. To see for myself, I decided to visit the Danforth Plant Science Center and the mesmerizing growth house that I described at the beginning of this chapter.

I made the trip early one autumn morning in 2017. As my plane dropped down out of the clouds on the approach to St. Louis, I looked out of my window and saw some of the world's richest farmland come into view: flat midcountry plains occupied by seemingly endless fields of different shades of green, with roads and railroad lines skirting the fields, and small towns and cities sitting at great distances from one another. The expanses feel oceanic. In the midst of it all sits St. Louis, a vast metropolitan island.

Agriculture plays an important role in the region's economy. St. Louis and its environs are home to the headquarters of some of the country's most important food companies—and, for that matter, to universities and research centers, including Washington University and the University of Missouri campuses, full of agricultural experts. The founders of the Danforth Center chose its location to maximize the kinds of synergies that would enable their mission "to improve the human condition through plant science."

I was met at the Danforth Center by one of its Distinguished Investigators, Dr. Elizabeth Kellogg, a plant biologist with an incredible breadth of interests and insights. Kellogg joined the Danforth Center in 2014 and holds appointments at both Washington University and the University of Missouri–St. Louis. She has devoted her career to studying cereals and their relatives in the grass family. Cereal crops have anchored human nutrition and human civilizations as far back as we can trace the history of agriculture.

Kellogg took me on a quick tour of the grounds, drawing my attention to the indigenous plants of the Missouri plains—a gently wild array of grasses, brush, and trees that had turned an autumn gold. The prairie plants came from local seed and are representative of the original prairie vegetation that was interlaced with forest in this part of Missouri before the Europeans arrived. Kellogg could not contain her delight as she described the Danforth Center project that had replaced the recently developed standard landscaping plants with plants reflecting the region's ecological history. At a slight distance stood a small village of greenhouses and, beyond that, meadows and undeveloped grassy areas interrupted by the headquarters of various agricultural companies.

Entering the main building, I was struck by its openness, which afforded a view all the way through the building to the greenhouses and fields beyond. I slowly realized that what was once a standard corporate headquarters layout had also been converted, like the grounds, to reflect a different approach to discovery. In addition to the entryway's view through the building and into the distance, it also opened directly onto a large gathering place that served as the dining and congregating center. Kellogg explained that the key organizational thesis of the Danforth Center rested on collaboration not only among the scientists, students, and technical staff in the building but also among the center, the universities, and the agriculture businesses in the area. Along with Kellogg, many of the Danforth scientists have appointments at the local universities; many also have active connections with industry partners. Every project I was introduced to during my visit, every facility I visited, every distant goal I was told about—all were described as group efforts. The spirit of collaboration animates the Danforth Center.

At the end of our tour, Kellogg took me to meet Dr. Jim Carrington, the president of the Danforth Center. We met in a conference room where Carrington and almost a dozen researchers had gathered to tell me about their work. Carrington, a celebrated plant molecular biologist, speaks with a level of precision that most of us achieve only in writing. He has an ability to reduce complicated ideas to their most important and most comprehensible basics.

Carrington began our discussion by introducing the challenge of rapidly escalating our food productivity. "If we expect to meet the world's needs anticipated in 2050 using today's methods and technologies on the farm," he said, "we would require additional farmland at least as large as the continents of Africa and South America combined." Since that's obviously not an option, Carrington and his team are searching for new ways to feed the world's population sustainably, which means using the many shared facilities at the Danforth Center to develop technologies and strategies for increasing crop productivity, in the United States and around the world, without growing the environmental footprint. It's a multifaceted, multidisciplinary effort, but underlying it all, Carrington explained, is one compelling idea: exploring the natural genetic variation that exists within plants today to help engineer better crops for tomorrow. We have a lot to learn, Carrington feels, from nature's genius. The genetic complexity of plants is a treasure trove of possibilities from which we can derive better plant variants, through both genetic modification and targeted breeding. But we can only do that fast enough if we can employ a selection process that tests many plants over their life course.

Carrington then turned the floor over to the scientists in the room, who gave me a whirlwind introduction to the many different projects they run to evaluate plant variation. One scientist explores the possibilities of using RNA to inhibit gene expression in plants. It's an alluring idea: rather than adding genes that express some trait, like a toxin that kills invading insects (Bt, for example), we might be able to deploy an inhibitory RNA to turn off specific genes and therefore reveal the desired trait. Another scientist adapts X-ray technologies used to detect metal fatigue in the automotive and aircraft industries to image and measure the growth of a plant's root system. This has critical importance for root and tuber crops such as cassava and potatoes, where the food products develop below the soil surface. Other plants, such as soybeans and peas, accumulate nitrogen in their roots, which means that they can supplement the nitrogen content of soil; this is one important reason for crop rotations that include nitrogen-fixing plants like clover. Much of the current demand for nitrogen is met by applying exogenous nitrogen in fertilizer, but if the Danforth scientists can use root imaging technologies to understand this process better, they may be able to amplify some plants' ability to accumulate nitrogen on their own, which would allow farmers to lower their demand for nitrogen fertilizer.

After our meeting, I headed off with two Danforth scientists, Becky Bart and Nigel Taylor, to visit some of the specialized facilities at the Danforth Center—namely, facilities devoted to phenotyping, microscopy, bioinformatics, proteomics, mass spectrometry, and plant tissue culture. We peeked into laboratory suites and visited one small part of the Danforth greenhouse complex of forty-three experimental stations and eighty-four plant growth chambers and rooms.

Then we came to the growth house with its unforgettable dancing plants, choreographed to travel along conveyor belts in complex pathways. The most important feature of the growth house, Bart and Taylor explained to me, is that it gives scientists the ability to computationally control manipulations at the level of individual plants. That affords lots of experimental variation and allows several experiments to run simultaneously. In a single experimental run that can last as long as six weeks, several different varieties of a plant species might be tested under identical conditions, or a single variety of plant might be tested for response to different watering or fertilizing conditions. Scientists can mix and match these paradigms as they see fit. A single experimental period might test five different drought conditions on a handful of well-characterized, genotypically different plants, for example, to determine which genotype produces the best phenotype. This is a key question for plant development today, and the faster this kind of experiment can be conducted, the faster agricultural productivity will increase. Because drought poses perhaps the most significant challenge for agriculture in the years ahead, many of the experiments conducted in the growth house test plants for drought resistance.

As we left the growth house, Bart and Taylor showed me one of Danforth Center's new acquisitions: an industrial robot arm they've named Grace. Standing about ten feet tall, Grace carries cameras that can fully photograph a plant from many angles, using different kinds of cameras to capture a variety of properties over the course of a plant's development. The robot is secured onto a platform that eliminates all vibration, so it can capture images at the highest possible level of resolution. Grace will soon become one of the many stops that plants make on their way through the Danforth Center's growth house.

Imaging is the key technology in the growth house. The array of cameras deployed there includes standard-light cameras that can capture plant size and leaf patterns, fluorescent-light cameras that can monitor plants' ability to absorb light and measure their stress responses, and infrared-light cameras that can detect water content. But imaging is just the first step. All of the images taken by these various cameras have to then be reduced, using state-of-the-art computers, to numerical representations—vast amounts of data that those computers then have to analyze.

Today, indoor phenotyping facilities like the Danforth Center's growth house make it possible to gather the data on hundreds of plants a day. That's a vast improvement compared to doing the work manually, which requires roughly an hour or more per plant depending on the phenotype. But the volume of data that these new indoor facilities can gather is so large that processing it requires not only state-of-the-art computers but also sophisticated software that has been written specifically for image processing. The Danforth scientists have created their own software for this purpose, called PlantCV, and have made it available as an open-source resource to researchers around the world. Collaboration, as always at Danforth, is the name of the game.

So far, so good, but the ultimate goal is to study plants as they grow in the field. To that end, Todd Mockler, another of Danforth Center's Distinguished Investigators, leads a large consortium that is developing a field-testing facility in Maricopa, Arizona. The Maricopa facility is a bit like the Danforth Center's growth house on steroids. The project, TERRA-REF (Transportation Energy Resources from Renewable Agriculture Phenotyping Reference Platform), a fifteen-plus-member collaboration situated at the University of Arizona's Maricopa Agricultural Center and the Department of Agriculture's Arid Land Research Station, has transformed a field measuring 20 by 200 meters into what may be the largest in-field phenotyping facility in the world.

The field can support as many as 80,000 individual plants, which might represent as many as 400 different genotypic varieties. The field is surveyed by a complex instrument system, called the Lemnatec Field Scanalyzer, which includes a metal gantry, mounted on steel rails, that glides back and forth along the full length of the field. The gantry carries an array of sophisticated cameras that gather all sorts of phenotypic data about the plants growing there—namely, size, growth rate, leaf patterns, color, shape, crop yield, disease tolerance, and water retention. Most importantly, these data are recorded for each individual plant daily over the course of its growth and maturation. Like the growth house, the Maricopa facility uses new imaging and new computational capabilities to provide plant-by-plant analysis at very high resolution to tackle the daunting problem of a large field planted with many varieties.

The high-resolution acquisition of data that represents plant traits, as well as environmental conditions, makes the promise of high-throughput, in-field phenotyping possible. The Maricopa system, Mockler told me, is designed to follow the development of the individual plants in the field for as long as two months. Once the vast treasure trove of data is collected, the scientists who run the facility analyze it using advanced computational methods, including machine learning and other kinds of artificial intelligence—all of which allow them to identify the few very best plant varieties for future propagation. For some plants, the facility has already developed the ability to predict crop yield from a plant's phenotype at thirty days. The work that the Maricopa scientists are doing is also providing foundational materials for the further identification of key genes and biological processes for next-generation genetic modifications.

o

The genotyping and phenotyping revolutions have already produced remarkable advances in the early years of this century. According to a recent report published by the US Department of Agriculture, more than 90 percent of the acres farmed in the

United States for corn and cotton in 2017 used genetically modified variants, up dramatically from 2000, when the comparable figures were less than 60 percent for cotton and less than 30 percent for corn. But even these dramatic advances fall short of what we need if we are to feed the growing population around the world in the decades ahead.

Consider cassava, a crop with enormous significance in the developing world. It grows in poor soils and resists drought. Its roots produce large tubers, which have become among the most important staple foods for more than half a billion people throughout the tropics. Cassava is grown mostly by subsistence farmers on small plots for on-farm consumption and to sell in local markets. Cassava's economic profile does not attract the kind of investment from the agriculture industry that has driven yield increases in commodity crops such as corn, soybeans, and cotton.

The Danforth Center is one of the places with a research program to improve cassava through genetic modification and phenotypic selection. Nigel Taylor, one of the scientists I met earlier, explained to me that the work began in response to the Cassava brown streak disease (CBSD), an insect-transmitted viral disease that in recent years has suppressed yields of this critical crop and threatened food and economic security for farmers in East and Central Africa. The most promising approach to combatting the disease, Taylor said, seemed likely to be gene suppression. When the plant is genetically modified to express a gene sequence from the viruses that cause CBSD, the gene triggers the plant's intrinsic disease recognition mechanism, causing it to power up its defense systems. Like immunization, the plant is forearmed, which makes it able to attack the pathogen immediately after an infection and before it can establish the disease.

Taylor told me about how a team of scientists carried out initial work on the project in Danforth Center greenhouses. Their work was part of the VIRCA (Virus Resistant Cassava for Africa) and VIRCA Plus projects, which engage scientists and government organizations in Uganda and Kenya, as well as the Danforth Center, he told me. Starting in 2008, they first isolated the viruses that cause CBSD in the field so that they could identify the best gene sequences for triggering disease resistance. They then introduced these sequences into the genome of hundreds of cassava plants and determined which of the plants stably and strongly expressed the introduced gene and best resisted disease. Eventually, they identified twenty-five promising cassava plant lines and then, working with the international collaboration, planted these lines under confined field tests in Uganda at locations where CBSD is especially severe.

After the first round of testing, the consortium selected about half a dozen lines for repeat testing, which they conducted at different locations in Uganda and Kenya, over successive planting cycles. The best plants needed not only to be resistant to CBSD but also to produce yields of the quantity and quality required by farmers. That narrowed their selection to two lines, which are now undergoing the further assessment required by regulatory bodies in both countries. The evaluations follow international standards for food, feed, and environmental safety, similar to the evaluations carried out hundreds of times to assess and approve other genetically modified crops such as corn, soybeans, and cotton. If all goes well, farmers in East Africa may have access to the new variants free of charge as soon as 2023.

High-throughput phenotyping may allow us to modify cassava in other helpful ways. Root phenotyping is one of the technologies under development, using X-ray technology to image roots as they grow underground. This would be a critical advance. To date, studies of plant phenotypes have mostly assayed shoots and leaves, which can be visualized easily. For root and tuber crops such as cassava, shoots and leaves provide a poor proxy for determining the quality and quantity of the food product or for monitoring a root-destroying virus such as CBSD. Developing root phenotyping for cassava could permit far more accurate monitoring of CBSD and other yield-impacting factors and facilitate the development of new strategies to combat disease.

Genetic modification of cassava also offers the possibility of increasing its nutritional value. Cassava storage roots are an excellent source of calories but contain very low levels of essential micronutrients. In many countries where cassava represents one of the dietary staples, a large fraction of children and women are anemic, according to the World Health Organization. Unmodified cassava can't help with this problem because it contains zinc and iron at levels too low to prevent malnutrition. But the international team within the VIRCA Plus project has developed new varieties of the plant that accumulate significantly more iron and zinc in their tuberous roots, holding the promise of amplifying delivery of these elements for undernourished populations. The consortium's long-term goal is a nutritionally enhanced, disease-resistant variety of cassava.

To improve cassava and other crops for both the developing and the developed world, we will certainly draw on our rapidly growing understanding of individual genes and the complex mechanisms that regulate their expression and their stability. But given what we're learning today about the multigenic regulation of traits, we will also have to continue to draw increasingly on phenotyping. This will require more of the high-throughput techniques and technologies that I saw at the Danforth Center: accurate, rapid methods for measuring and recording plant traits, and computational tools for analyzing the vast amounts of data that those methods will deliver to us.

We don't know—yet—the genes that determine all of the complex traits that we seek in our crops. But as a result of the astonishingly powerful ways in which biology and engineering are converging in agriculture, we can be sure that the day will soon come when we do. That thought gave me great hope as my plane took off from St. Louis. We are up to the challenge, I thought, as vistas of bountiful farmland opened up before us: we have a path to technologies that will help us provide affordable, nutritious food to more than 9.5 billion people all over the planet.

7

CHEATING MALTHUS, ONCE AGAIN

Making Convergence Happen Faster

In 1937, seven years into his tenure as the president of MIT, Karl Taylor Compton wrote a delightful article celebrating the fortieth anniversary of the discovery of the electron, titled "The Electron: Its Intellectual and Social Significance." The discovery, Compton reminded his readers, had been made in 1897 by the physicist J. J. Thomson, who had not only identified the particle but had also determined that it alone was responsible for electric currents. Physicists at the time thought that the atom was the smallest particle in existence—the indivisible building block of all matter. But Thomson's discovery shattered that idea. Thomson won the Nobel Prize in 1906 for the discovery. But five years later, Compton noted, some physicists still refused to accept it and its revolutionary implications.

That stance became impossible to maintain in the years that followed, as Thomson's discovery made possible an expanding array of almost magical "electronic" technologies: radio that could send transatlantic messages; long-distance telephone service, which for the first time allowed real-time conversations at great distances;

photoelectric devices, used to detect movement (and open doors) and to replace photographic film in cameras and telescopes; movies with soundtracks; and more. After reviewing this set of technologies, Compton declared the electron to be "the most versatile tool ever utilized," but he then went on to suggest that the full impact of its discovery had yet to be felt. He was right, of course. Neither he nor anybody else in 1937 could foresee what the electronics industry would become or how that industry would give rise to the technology transformations of the twentieth century: the development of the computer and information industries and our now ubiquitously digitally enabled world.

Thomson's discovery opened up the door to not just new technologies but also new discoveries. Following his lead, scientists threw themselves into the study of the subatomic world and soon discovered the neutron and the proton, the remaining components of the atom. And they didn't stop there. We now know that neutrons and protons themselves have component parts: namely, quarks and gluons.

The discovery of subatomic particles opened up the field of nuclear physics, which has led to all sorts of revolutionary technologies, among them electricity generation from nuclear power plants and remarkable imaging capabilities from nuclear medicine. But as is often the case for new discoveries, scientists at the time of these discoveries had little conception of what their work would ultimately engender. Ernest Rutherford, widely acknowledged as the father of nuclear physics, discovered the proton in 1909 and described the atomic nucleus in 1911. But even he did not foresee the practical implications of his work, asserting in 1933 that "anyone who expects a source of power from the transformation of these atoms is talking moonshine." Yet less than twenty years later, in 1951, the United States demonstrated the production of electricity from a nuclear power plant at the Idaho National Laboratory.

Rutherford's inability to anticipate the practical uses of even his own discoveries is a common story. Try as we might, we rarely accurately predict which insights into the fundamental phenomena of our universe will provide the basis for later technologies. Nonetheless, fundamental discoveries are necessary for the development of new technologies that can bring great human and economic benefits. In a perhaps apocryphal story from the 1850s, the British Chancellor of the Exchequer, William Gladstone, challenged Michael Faraday's groundbreaking discoveries of the behavior of electricity and electromagnetism, questioning whether they had any practical use. Faraday conceded that he could not describe a specific application, but that didn't diminish his confidence in the potential of what he had discovered. "Why, sir," he is said to have told Gladstone, "there is every probability that you will soon be able to tax it."

In the more recent past, governments have recognized that to benefit from economic advances fueled by new technologies, they must invest in fundamental research. The economic return on fundamental research has a too distant and too uncertain horizon for standard investments, so governments have shouldered the responsibility for planting the seeds of future economic returns, providing federal funds for early research. Countries that have made these investments have realized outsized returns in industrial and economic growth.

In the introduction to this book I noted that roughly a hundred years ago the discovery of the electron, X-rays, and radioactivity put at our disposal for the first time a basic "parts list" for the physical world and that this made it possible for an innovative generation of engineers to create remarkable new electronic tools and technologies. Far earlier than most, Karl Taylor Compton recognized that this convergence of physics and engineering represented the beginning of a new era of scientific innovation, which today we might call Convergence 1.0, and he did everything he could during his career, at MIT and elsewhere, to encourage interdisciplinary collaborations that would help maximize its potential. It's hard to overstate how profoundly Convergence 1.0 transformed our world. The digital technologies and computation it made possible are now such an essential part of our lives that we take them for granted.

Today, as I have noted, we have at our disposal another "parts list" from the biological world. With that list in hand, we stand on the verge of another convergence with engineering that promises to revolutionize our lives yet again. I've offered a few glimpses of exciting new tools and technologies that this convergence—Convergence 2.0—is already making possible: virus-made batteries, protein-based water filters, nanoparticles that can help detect and cure cancer, brain-powered prostheses, and computer-mediated rapid crop selection. These technologies, along with many others in development and still others that we cannot yet even imagine, offer us the prospect of a safer, healthier, and cleaner world.

The possibilities are thrilling. As Aquaporin A/S's founder and CEO, Peter Holme Jensen, told me, we may soon solve many of our problems by "just using nature's genius." We may soon be able to enlist viruses not only to cleanly and efficiently produce batteries, as Angela Belcher is doing, but also to take on jobs that I have not addressed, among them turning methane into ethylene (the key component of plastic bags, bottles, and boxes) or catalyzing the fixation of nitrogen (an energy-intensive step necessary to produce the mass quantities of fertilizer necessary to feed Earth's growing population). We may soon be able to enlist nanoparticles not only to detect and treat cancer, as Sangeeta Bhatia is doing, but also to help reverse climate change by capturing carbon dioxide from the air and then turning it into useful industrial and commercial products, such as coating products that can render just about any surface self-cleaning and water-repelling. We may soon be able to harness the power of plants to light our homes as well as harvest enough energy from natural sources—for example, the sun, the wind, and the tides—to satisfy our energy needs and end our reliance on fossil fuels.

But none of this is inevitable. Fostering Convergence 2.0 will require the financial commitment, interdisciplinary collaboration, and political will that made Convergence 1.0 so successful. It will require significant new investments in fundamental research, the creation of sufficient, long-term capital flows to foster new industries, and immigration policies that reinvigorate the nation's welcome to the best and the brightest from around the world.

Convergence-minded policies and practices don't come into being by themselves. They didn't exist when Thomson discovered the electron in 1897; or when Rutherford described the atomic nucleus in 1911; or when Karl Taylor Compton, well into his presidency at MIT, wrote his article celebrating the electron in 1937. By the time of Compton's article, the basic discoveries necessary to enable Convergence 1.0 had already been made, but the United States was still recovering from the Great Depression and had not invested sufficiently to deliver the full potential of Convergence 1.0 products and industries. Unemployment stood at over 14 percent and would rise to nearly 20 percent the following year, and manufacturing output was in decline. Few people could have predicted that the country would emerge just a few decades later as the world's technological, educational, and economic superpower.

What made the difference was the Second World War and the major multidisciplinary efforts that the United States launched with other nations to develop such war-winning technologies as radar, sonar, new forms of computing, and the atomic bomb—all Convergence 1.0 technologies.

Following the war, the nation's commitment to continue federal investments in research—as championed by Vannevar Bush, the principle architect of the country's wartime and peacetime technology programs—fueled the postwar industrial and economic growth that propelled the United States into a position of world leadership. Nations around the world have learned the lessons of that success and are racing to replicate its ingredients: ambitious federal research and development (R&D) investment strategies, world-class research universities, welcoming immigration policies, and forward-looking industrial models. Almost every week during my MIT presidency, I received a visitor from a country aspiring to reproduce the economic miracle that America managed to engineer in the twentieth century. These countries had ambitious plans—and today, more than ever, they are acting on them, increasing their investments, and putting policies in place to develop the technologies of the future.

o

Thanks to the privileged view of the scientific and technological future granted to me as the president of MIT, I have seen the promise of Convergence 2.0, which has the potential to transform the twenty-first century every bit as dramatically as Convergence 1.0 transformed the twentieth. Perhaps even more dramatically, in fact, given that the tools and technologies we are now poised to develop have the potential to directly combat many of the problems that threaten us most gravely as a species and as a planet. But what I wonder today is whether the United States can mobilize itself to lead the way for Convergence 2.0 in the manner it did for Convergence 1.0 and, moreover, whether it can do this without the unfortunate accelerant of war. This strikes me as one of the great political questions of our time—and it is a question that we must try to answer.

The convergence of biology and engineering provides a very good reason to hope that we can once again avoid the gloomy future that in 1798 Thomas Malthus described to be our fate: a future of inevitable war, famine, and pestilence. Already, as I've tried to show in this book, we have game-changing new technologies at our disposal. So we must now ask the critical question: How can we speed these technologies into use? And how can we create conditions that will allow us to bring more of them to light as quickly as possible?

At the most basic level, dialing up a few of the controls on our innovation dashboard would produce outsized gains. From our experience with Convergence 1.0 in the twentieth century, we already know what works: amplifying federal investment in fundamental research that encourages cross-disciplinary and cross-institutional projects and education; designing technology transfer practices that speed new ideas into the market; developing fiscal policies that encourage investments in long-cycle, capital-intensive industries; and implementing immigration policies that keep our research operations as the strongest possible magnets for the world's talent. Committing ourselves to these strategies is not an impossible ambition. But to unleash the full potential of Convergence 2.0, we will need to rethink the structure of our educational institutions and research laboratories, our funding agencies, and our financial policies, all of which currently impede discipline crossing. Let's briefly consider how we might do this.

o

None of the examples I have described would be possible without government funding. The sustained federal commitment to funding discovery research and technology development (R&D) enabled the United States to rise to its position of technological leadership in the twentieth century. Following the massive R&D investments that produced war-winning technologies for the Second World War, the country made a major new commitment at the federal level to continue to invest in research and development. The thesis, in the words of Vannevar Bush, was that "the lessons learned in the war-time application of science . . . can be profitably applied in peace."

By the mid-1960s, the level of federal investment in R&D had reached 2 percent of GDP. Percent of GDP is the best benchmark because it shows the level of the society's commitment to research. Disbursed by federal agencies dedicated to the idea of becoming better by becoming smarter, this funding brought new business ventures and even new industries into being. The effort paid off handsomely in the final decades of the century, with the explosive growth of the computer and information industries and all of the tools and technologies they enabled.

Shortsightedly, the US government has scaled back its investment in R&D, which has now declined to less than 1 percent of GDP. While private sector R&D as a percent of GDP has grown as federal R&D fell, it is not a substitute. Private and public R&D focus on different roles: in public R&D, federal funds support predominantly early research while the private, industry sector R&D funds focus predominantly on the development of research discoveries into marketplace products. We must have both; they are not interchangeable.

For a number of reasons, the decline in federal R&D funding has hit cross-disciplinary research particularly severely. When funding levels stagnate or drop, decisions on the allocation of resources become increasingly conservative, with support going toward more predictable possibilities rather than toward more uncertain new directions, like Convergence 2.0. Moreover, most of the federal research investment comes through the major research agencies—namely, the National Institutes of Health, the National Science Foundation, and the Departments of Energy and Defense—all of which are guided in their research targets by twentieth-century disciplinary boundaries. Having separate agencies dedicated to separate disciplines makes funding for cross-cutting Convergence 2.0 projects exceedingly difficult, if not impossible, to secure. In some cases, private philanthropy has stepped up to provide the funding needed to launch and sustain new cross-disciplinary approaches. According to the Science Philanthropy Alliance, private philanthropic funds for basic science research (of all kinds) amounted to around $2.3 billion in 2017. (This number is based on survey responses so is likely an underestimate.) Although this is only a fraction of the federal research commitment, it can help pave the way for new scientific directions. It was foundation funding, for example, that demonstrated the possibilities of a Convergence-based Brain Initiative.

Declining federal R&D investments have hit the physical sciences and engineering most severely—fields of critical importance for Convergence 2.0. The American Association for the Advancement of Science reported that between 1970 and 2017 funding for projects in these fields dropped by about 55 percent (relative to GDP). Even for biomedical research, federal expenditures on R&D, after doubling between 1996 and 2003, have declined significantly, dropping by nearly 22 percent in purchasing power from 2003 to 2017. Recently, Congress has raised National Institutes of

Health (NIH) funding levels, but that has generally not been the case for the physical sciences. If federal investments continue to stagnate or decline in the years ahead as Convergence 2.0 opportunities accelerate, the United States will not be able to embrace them or take its position as a global technology leader for granted.

The decline in federal research investments makes little sense on a national level, but it makes even less sense when considered in a global context. Between 1995 and 2015 many nations—adopting the strategy that delivered such remarkable economic success for the United States in the twentieth century—worked hard to increase their government and industrial investments in technology R&D. Some of the most dramatic increases came in China (now more than 2 percent of GDP), South Korea and Israel (both now more than 4 percent), and Japan (well over 3 percent). The magnitude of these countries' R&D investments now positions them to compete with the United States, where combined government and industrial R&D investment rests at about 2.8 percent of GDP. The country that once led the world in inventing technology's future, in other words, now risks lagging behind, especially given China's planned increases in investment.

The decline in federal research investments also makes little sense when considered against the benefits they can foster. Taking only one example, the National Institutes of Health, with an annual budget of around $37 billion for fiscal year 2018, funds the majority of discovery research in biology and medicine. The returns on that investment have been measured in a variety of ways. Considered as savings from disease prevention, for example, the Centers for Disease Control and Prevention (CDC) has estimated that for children born in 2009 alone, childhood vaccinations—many of which were developed with NIH funding—will save 42,000 lives, prevent 20 million cases of disease, and reduce health care costs by $13.5 billion. And thanks in large part to new medical insights and technologies funded by the NIH, we have witnessed a rise in the average life expectancy for Americans from less than seventy years in 1960 to more than seventy-eight years in 2015, an increase that has an estimated economic value of about $3.2 trillion per year. That's an astonishing return on investment.

We face another significant hurdle before we can make the most of the opportunities offered by Convergence 2.0. Federally funded research grants generally are based on the "solo adventurer" and "singular discipline" models, both of which are ill-adapted to broad-based, multidisciplinary, convergence-style collaborations.

Fortunately, the US government, recognizing the need for new funding models, has experimented with cross-disciplinary and cross-agency initiatives. One of the best known among them is the Human Genome Project, launched in 1990 by an international collaboration primarily funded by the US-based National Institutes of Health and Department of Energy and the UK-based Wellcome Trust, working in competition with a private sector team. Between 1990 and 2003, the project brought biologists, computer scientists, chemists, and technologists together to develop new methods for gene sequencing. Early successes included the first maps of the fly, mouse, and human genomes, providing insights into a host of biological processes. It also set the stage for the genetic analysis of disease, which has now allowed the identification of candidate genes for cancer, diabetes, schizophrenia, and more. Critical to these achievements was the development of new technologies for DNA sequencing, which dramatically drove down the cost. In 2001, sequencing a human genome cost over $100 million; today it costs less than $1,000. Thanks to the Human Genome Project, we have tools in hand to understand disease at an entirely new level, which will permit diagnoses and treatments targeted to an individual's unique genotype and specific disease subtype.

More recently, the National Nanotechnology Initiative (NNI), launched in 2000, has brought together twenty federal departments and agencies with the goal of accelerating progress on research and industrial applications at the nanoscale. NNI projects have ranged across disciplines, including the development of quantum dots (nanoscale semiconductors) for medical imaging, new compositions for battery electrodes, and nanomaterials that can extract hydrogen from water. And in 2013 the government launched the Brain Research through Advancing Innovative Neurotechnologies Initiative (BRAIN). A ten-year project, BRAIN brings together neurobiologists, engineers, and physical scientists working across three agencies to design new technologies to unravel the complexities of the mind and of the diseases that can destroy it. Among its goals is to miniaturize brain-computer interfaces, like those used by John Donoghue and others to record brain activity, and to produce higher-resolution maps of brain function. These advances promise to make the pioneering prosthetic technologies described in Chapter 5 accessible to many more of the people who need them. Comparable cross-disciplinary and cross-agency initiatives are now underway for Precision Medicine, the microbiome, and the Cancer Moonshot.

Today, despite their manifest successes, multidisciplinary and cross-agency research projects like these are still the exception rather than the rule. That will have to change if we are to make the most of Convergence 2.0.

o

We will need to make changes outside of the government sphere, too. Consider the way in which most universities today organize their faculties and academic departments—namely, by disciplines. In lots of obvious ways, this makes good sense. A chemistry department made up of formally trained chemists can organize the courses and experiences that, building from one set of ideas to the next, prepare students to become experts in chemistry. It can share research facilities and host seminars on topics of mutual interest. And the best such departments, in designing their curricula and creating research programs, determine the future of their fields.

Over time, however, departmental and disciplinary boundaries can become rigid and impermeable. Each discipline develops its own history, its own vocabulary, its own definition of problems and viable paths to discovery, and all of this discourages cross-disciplinary collaboration and understanding—precisely the ingredients necessary to accelerate scientific convergences of the sort I've described here.

During my time at MIT, we have worked hard to break down barriers to cross-disciplinary collaboration and understanding. When we created the Koch Institute for Integrative Cancer Research, for example, which brings together biologists, engineers, and clinicians, we began with the premise that each of the Institute's members would have to learn some of the language and the problem-solving approaches of the others. To that end, sessions like "Engineering Genius Bar," "Crossfire," and "The Doctor Is In" have given people a way to fill the gaps in their understanding. We didn't want the engineers to simply be called on as "service providers" when the biologists had reached a dead end.

This approach very quickly bore fruit. New collaborations spawned new insights. Just to give one example of dozens of the resulting new approaches, Professor Paula Hammond, a chemical engineer who pioneered layer-by-layer nanotechnology fabrication methods that she used to build energy storage devices, partnered with Professor Michael Yaffe, a physician and molecular biologist. Together they developed nanoparticles that deliver two anticancer drugs in a carefully timed one-two punch to amplify the effectiveness of chemotherapy.

I'm not proposing that we do away with departmental structures, which serve many important purposes. Nor am I proposing that we suddenly reorganize our departments under other names and with different purposes. Early in my presidency, when I was asked whether MIT needed to engage in this sort of renaming and reorganization, I responded no. I felt that we simply couldn't know what disciplines or directions would be most important in even a few decades. So instead of a departmental reorganization, we chose a different approach: we would draw on the history and strength of current disciplines to seed new interdisciplinary labs and centers and would try whenever possible to provide two homes for faculty, one in an academic department and a second in a research center. This was a model that MIT had already deployed after the Second World War as a way of continuing the campus-based multidisciplinary collaborations that had been so productive in the development of radar. Since then, several technology- or problem-focused research centers have fostered collaborations in the physical sciences and engineering. By expanding that model now to serve Convergence 2.0, we hoped to maintain our traditional disciplinary strengths while also pursuing new kinds of cross-disciplinary collaborations that would evolve—or dissolve—as they succeeded or failed in their missions. The approach has succeeded beyond our expectations. Many other campuses have experimented similarly with new organizational structures to bring different disciplines together. Promising models include the International Institute for Nanotechnology at Northwestern University; the Institute for Regenerative Engineering at the University of Connecticut School of Medicine; the Wyss Institute and Centers at Harvard, ETH Zurich, and Geneva's Campus Biotech; and the joint bioengineering program at UC-Berkeley and UC-San Francisco.

We can do more to tune our educational systems to encourage not just discovery but also innovation and new industries. Few graduate programs provide on-ramps to industry, even though an increasing number of PhD recipients in the sciences and engineering join or start companies after graduating. This is changing in some quarters. Northeastern University in Boston, for example, recently launched an "experiential PhD program" in collaboration with GlaxoSmithKline, designed to give students concurrent experiences from academic and industry perspectives. And Danish universities now offer industrial PhDs and postdoctoral positions, funded jointly by the government and industry. We will need more of these kinds of programs and more creative thinking about graduate education if we want to encourage the development of Convergence 2.0 products.

o

Convergence 1.0 taught Americans that federal R&D investments can help catalyze the development of new products, new businesses, and even new industries. But to reap the full benefit of these kinds of federal investments, we will also need to accelerate the pace of moving ideas out of the lab and into marketplace products, something that benefits from new relationships among government, academia, and industry. Before 1980, for example, the ownership of the patent rights resulting from federally funded research belonged to the federal government, but federal agencies did not have appropriate mechanisms or incentives to drive the development of research products into the market. As a consequence, the nation did not reap the potential economic benefits from its massive research investments.

All that changed in 1980 with the passage of the Bayh-Dole Act, which aimed to accelerate the transfer of research discoveries into commercial products. The act assigned the ownership of intellectual property developed with federal funds to the organizations that had carried out the research—universities, nonprofit organizations, and small businesses. Transferring patent ownership to these organizations gave them a strong economic incentive to develop their discoveries into applications that could be taken to market. One measure of the success of Bayh-Dole is the dramatic increase in the number of patents issued to US academic institutions, from less than 500 in 1980 to just over 2,000 in 1996 and almost 7,000 in 2016.

Stanford and MIT are often cited as among the most effective universities at technology transfer and usually rank among the top few in the US Patent and Trade Office's annual report on university patenting activity. Both also rank among the top few in spinning off start-up firms. Part of that success is attributable to excellent engineering programs, which tend to be product oriented. But, in addition, both schools' history led them to develop cultures and policies that facilitate engaging with industry. Both were founded in the second half of the nineteenth century. MIT's founding mission included engaging with practical problems in laboratories where students would be "learning by doing" and in curricula that included "mechanical training" (which today we'd call engineering) and stressed the value of "useful knowledge." Stanford's rise began in the 1950s through systematic research connections with the electronics and early computing industries that were then starting up in its Silicon Valley neighborhood. Both schools were positioned, in other words, to help accelerate the industrialization of America and, in many ways, they both embedded technology transfer into their DNA.

Today, both schools' technology transfer goals prioritize moving research products into industry development—as many as possible, as quickly as possible. From experience, they know that most new products and new ventures fail, and they recognize that it's in everyone's interest to make it relatively easy to transfer technology simply and quickly. To facilitate that, their technology transfer offices include people with industry experience so that the offices can understand the issues from both sides. In addition, both schools benefit from a long history of technology transfer, which provides a library of prior experience and an invaluable set of established industry relationships. They view building partnerships as one of their central responsibilities. By contrast, some organizations follow a model that prioritizes financial return, which can slow the flow of technology into industry and may hinder the kind of long-standing relationships that can have benefits in the longer term.

With technology transfer as a central part of their mission, Stanford and MIT have both deeply engaged with their local economic ecosystems. The vibrant industry hubs surrounding Stanford (Silicon Valley) and MIT (Kendall Square) reflect that philosophy, and both the universities and the local innovation hubs have enjoyed the economic and societal benefits of mutually beneficial, synergistic relationships. Many other institutions have followed suit to help build their local innovation ecosystem and accelerate the translation of research-based discoveries into products for the marketplace.

As with Convergence 1.0, so with Convergence 2.0—long-term investment will be key to delivering on its promise.

In the competitive world of new-company financing, investors often prefer software to what's known as "hard tech" or "tough tech." Software includes everything from social media platforms to online search algorithms to video games. It's relatively cheap and fast to develop, and investments in it can pay off extremely well, at least in the short term—ask anybody who a decade ago put money in Facebook or Google. Hard tech is different. It includes physical tools and technologies that require years and extensive infrastructure to research and develop and still more years of production design and scale-up to get ready for market. All the Convergence 2.0 products I've described here are hard tech—the virus-based batteries, the protein-based water filters, the nanoparticle-based cancer detection systems, the brain-enabled prostheses, and the computational-mediated selection of new crops. So, too, are such Convergence 1.0 products as next-generation jet engines and nuclear reactors. These physical objects can transform how we live, but getting them to market is hard: they require investors who take the long view, who recognize the promise of next-generation products, and who are willing to wait patiently for a long time before a payoff arrives.

The story of Aquaporin A/S provides a good example of the challenges that companies face in developing Convergence 2.0 technologies and of the benefit that visionary investors can provide. The company had to invent new methods to make industrial quantities of aquaporin protein for their water filters. They had to first come up with a new protein-manufacturing process for membrane proteins because the standard biopharmaceutical process produces proteins in solution. They then had to develop an entirely new production facility for their membranes. Considering the difficulty of mobilizing sufficient, and sufficiently patient, capital, I asked Claus Hélix-Nielsen how the company planned to navigate the time and expense required to bring the filters to the market. Hélix-Nielsen explained that the major investors in Aquaporin A/S, public and private Danish and Chinese entities, had a very long view of the return on their investments. "If the Aquaporin Inside™ technology works," he told me, describing their investment thesis, "it will indeed enjoy commercial success and, even more importantly, will provide a world-sustaining advance for water purification."

If we want to overcome the challenges we face in unlocking the potential of Convergence 2.0, we will have to do more to persuade investors to think along these lines. We will need to enact policies that encourage investment in long-cycle, capital-intensive industries with the specific goal of encouraging the development of Convergence 2.0 and other hard-tech products. One potentially promising way for the government to create such incentives has been proposed by Larry Fink, the chairman and CEO of the investment group BlackRock. The government, Fink suggests, should offer tax benefits that grow proportionally with the length of time of an investment. It's a sensible idea that would encourage capital to flow to hard-tech companies to the benefit of the American economy and lives around the world. We need to come up with more ideas like it—and then act on them.

We also need to reaffirm what we already know: that immigration is one of the powerful drivers of innovation in the United States. Consider this remarkable sampling of successful American companies with first- or second-generation immigrants as founders: Apple, Google, Amazon, Oracle, IBM, Intel, eBay, Tesla, Boston Scientific, and 3M. In 2017, nearly half of the founders of the Fortune 500 companies were first- or second-generation Americans, and well more than a third of all new graduate students in American departments of science, mathematics, and engineering (which are among the most competitive programs in the world) came from outside of this country. Young people can now travel anywhere in the world to pursue their dreams, and if we want to continue to be a powerful magnet for global entrepreneurial talent, we will need to create easier immigration paths for the adventurers who want to help start new companies and new industries. Student and worker visa programs should be expanded and paths to citizenship made easier. Unfortunately, this is not what we're doing. Recent proposals to limit immigration and reduce access to H1B visas (the route of entry for many scientists and engineers) are already starting to discourage many potential immigrants from venturing to the United States. International student enrollment in graduate programs flattened in 2016, and in the fall of 2017 the number of new international students enrolling in US colleges and universities declined for the first time since the 2001 terrorist attacks on New York City. This is not a trend that will keep us in the lead in a competitive global economy.

o

The small set of remarkable technologies I've described in this book are only a few examples of the kinds of new possibilities that could save us from the impending crises of too little energy, water, medicine, and food as we anticipate a global population of more than 9.5 billion by 2050. However, delivering on these kinds of exciting technological possibilities will rest in the hands and minds of the next generation.

My hope for a better future rests with as much confidence on the next generation as on the technologies themselves. Few conversations have inspired me more than those I've had with students at MIT and elsewhere. The young people I've spoken with understand with astonishing clarity the world's pressing challenges, and they have a passion to help find solutions to them. As president, at the start of every fall semester I would walk the MIT campus, asking the newly arriving freshmen to tell me about their plans and aspirations. Their answers surprised me: rather than describing a plan to major in biology or mechanical engineering or economics, most of their enthusiasm focused on our newest programs in energy and bioengineering. That enthusiasm gave much needed wind to our sails as we launched the MIT Energy Initiative and the several Convergence 2.0 institutes and programs.

Students everywhere want to know, "How can I help discover solutions?" The questions for our educational institutions are: How do we provide opportunities for students to engage in problem-based, purpose-driven activities? How do we get them involved in curing cancer and building sustainable energy solutions? How do we do that while equipping them with the critically important discipline-based knowledge that they'll need to invent the new technologies of their dreams?

The conundrum for students and educators is that our students have a lot to learn if they want to make real contributions to solving the world's difficult problems. They need a strong disciplinary foundation to productively apply their knowledge and know-how to real-world problems. But who has the patience to wait? MIT's answer has been to organize the undergraduate experience to include both disciplinary study and problem-based activities. As one example, when we launched the MIT Energy Initiative, a small army of undergraduates told us they wanted to get to work right now on designing a path to a sustainable energy future. To help them along, we decided to start a new undergraduate minor in energy—but not a new major in energy. If our graduates were to make important contributions as energy professionals, not only would they need deep disciplinary knowledge, but they'd also be better able to contribute if they could understand energy challenges from the perspective of a host of other disciplines. Nuclear engineers who understand the physics, economics, and politics of nuclear power can bring a breadth of perspective to the challenge of designing, building, and operating new nuclear power plants.

As important as official minors and majors are in education, students benefit enormously from participating in research. The post–World War II decision to embed most of the US scientific research enterprise within institutions of higher education has catalyzed productive collaborations between new and seasoned scholars and between the discovery and the transmission of knowledge. In research labs our students experience firsthand how the lessons they've learned in the classroom become the tools for finding problem-based, purpose-driven solutions. When an undergraduate cannot contain her excitement as she describes her research project on new chemical compositions for a cheaper, more efficient, environmentally benign battery, I know that we are succeeding as educators. And when an undergraduate devising new nanoparticles to deliver chemotherapy tells me about his brother's cancer diagnosis, I know that he has joined those marching down the path to a better future for all of us.

o

From my current perch in the Koch Institute for Integrative Cancer Research, I see every day the power of Convergence 2.0. Among my colleagues are the Nobel Prize–winning biologist Phillip Sharp and the world-renowned engineering entrepreneur Robert Langer, both fervent advocates of Convergence 2.0. The faculty and students come from all over the world and bring an incredible range of disciplinary expertise. They are astonishingly diverse, from any background, any nation, any phenotype or genotype. And yet, in the face of cancer's daunting threat, they share a commitment to crossing whatever boundaries they might encounter together. Bringing people together around a shared ambition amplifies their impact. Our institutions have provided the kind of catalytic environments where people and resources magnify individual talents in the service of a larger, shared purpose. Can we inspire this nation and all nations to invent a new path that mitigates the threats of drowning in rising seas, thirsting for want of clean water, dying prematurely from undiagnosed and untreated disease, living impeded by disability, and suffering political instability precipitated by insufficient affordable food?

I grew up in the shadow of Sputnik. But I didn't experience that historical moment as terrifying. Instead, I saw in it a bright beacon of hope: science and engineering could get us to the moon and beyond! That beacon opened for me the path to becoming a scientist, to studying how the brain assembles itself, to rethinking how scholars and researchers can collaborate across disciplines, and to designing innovative approaches to cross-disciplinary research at Yale and MIT to better serve the world. The path I've traveled, side by side with so many other scientists and engineers, has been enormously enlightening and rewarding. But there's still far to go. The future looms. We will face daunting challenges in the coming century, and to overcome them we will need to summon up a shared ambition and shared commitment that are every bit as powerful as the one that we summoned up to win the Second World War. But this time, I fervently hope, we will be motivated not by the threat of war but the promise of peace.

ACKNOWLEDGMENTS

After I stepped down as MIT's president, at the end of the 2012 academic year, I spent a postsabbatical year as a guest of the Belfer Center for Science and International Affairs, at the Harvard Kennedy School of Government. Having the Belfer Center as my home for the year afforded me a welcome opportunity to reflect on my presidency at MIT and on the Institute's history. During that year, my thoughts returned again and again to the idea of convergence that animates this book. Many others had grasped the possibilities of the idea well before I did, and I benefited greatly from their insights: Tom Magnanti, who was the dean of MIT's School of Engineering when I joined the Institute; Tyler Jacks, Robert Langer, and Phillip Sharp, the architects behind the Koch Institute for Integrative Cancer Research; Hansjörg Wyss, who put convergence concepts into clinical products decades ago; Ernie Moniz and Bob Armstrong, founders of the MIT Energy Initiative; and Bruce Walker and Susan and Terry Ragon, founders of the Ragon Institute. These and many others have helped accelerate convergence as a model for discovery and its seamless translation into real-world technologies.

At the end of my sabbatical year, I was invited to deliver the Edwin L. Godkin Lecture (an annual event at the Kennedy School). There, I gave a talk titled, "The Twenty-First Century's Technology Story: The Convergence of Biology with Engineering and the Physical Sciences." Following the talk, Graham Allison, then the director of the Belfer Center, forcefully suggested to me, in his inimitably encouraging style, that I share my story with a broader audience by writing a book. Others soon gave me similar encouragement, among them Suzanne Berger, William B. Bonvillian, Robert Putnam, and Phillip Sharp. Phil, in particular, made me understand how much more compelling the book might be if I could tell not only the stories of the new technologies that hold such great promise for a better future but also the stories of the remarkable people who were inventing them. That's the approach I adopted, and I give Phil full credit for the idea.

The stories I've told in this book emerged from conversations with dozens of scientists, engineers, social scientists, humanists, and entrepreneurs, all of whom shared their time and discoveries, and their dreams, with me. And then, if that wasn't enough, many of them reviewed early drafts of my chapters. Their patience, generosity, and good humor made writing this book among the most delightful educational experiences of my life. Selecting only a few examples from dozens as compelling and as promising as those I've profiled here was one of the most difficult decisions. Those decisions, along with whatever remaining errors and omissions, of course, are entirely my own.

I owe debts of gratitude to those who educated me as I researched and wrote: to Angela Belcher and her students, particularly Alan Ransil; to Peter Agre, who regaled me with touching and comical stories of his Aquaporin Adventure; to Claus Hélix-Nielsen and Peter Holme Jensen, at Aquaporin A/S; to Sangeeta Bhatia and her students, particularly Ester Kwon, Jaideep Dudani, and Tarek Fadel; to John Donoghue, a remarkable colleague over many decades, and his collaborator Leigh Hochberg; to Hugh Herr and Jim Ewing, true pioneers who have turned their own personal challenges into new technologies that improve the lives of others; to the wonderfully gracious leadership team at Össur, Hildur Einarsdóttir, Gunnar Eiríksson, Kim De Roy, David Langlois, and Magnús Oddsson, and my introduction to the company by Kristín Ingólfsdóttir; and to my hosts at the Donald Danforth Plant Science Center, particularly Elizabeth (Toby) Kellogg, who served as insightful guide and teacher, and James Carrington, Becky Bart, Mindy Darnell, Noah Fahlgren, and Nigel Taylor, all of whom read drafts of the material. In addition, I thank Susan Rundell Singer for suggesting high-throughput phenotyping as the key technology for the agriculture chapter and for her perpetual good ideas and careful review of the chapter, and Deborah Fitzgerald for sharing her historical perspective on the progress of agriculture. Barbara Schaal introduced me to the Donald Danforth Plant Science Center and has generously shared her wisdom on many matters. The historical and policy reflections in the introductory and concluding chapters benefited from many conversation partners, including Bill Bonvillian, Marc Kastner, Lesley Millar-Nicholson, Bill Aulet, Ed Roberts, Al Oppenheim, and MIT archivists Tom Rosko, Myles Crowley, and Nora Murphy. Geri Malandra, Bob Millard, and Lisa Schwarz reviewed early versions of the book and provided enormously helpful suggestions for how to make the material more understandable. These people and innumerable others, adventurers all, have become my heroes—and they have also become my friends, a wonderful and unexpected bonus.

This is my first book for a general audience, and I have received an education here, too. Toby Lester, with whom I developed the book from its initial proposal through every chapter, gave me perpetual encouragement. Toby understands how to turn a wooden recitation of facts into a compelling story and was an inspiring writing partner. My agent Rafe Sagalyn (ICM) introduced me to the behind-the-scenes world of book publishing and the still robust community of readers; he has been a constant guide in the process of turning ideas into a book. The book's week-by-week progress was immeasurably enhanced by the intellect, energy, and enthusiasm of my research assistant and thought-partner, Nabiha Saklayen. Erin Dahlstrom's meticulous referencing and fact-checking made these final, demanding details also a joy-filled experience. Luk Cox and Idoya Lahortiga at somersault18:24 created figures that brilliantly capture the essence of complex concepts. Quynh Do and John Glusman, my editors at Norton, provided expert guidance and encouragement from start to finish.

I am inexpressibly grateful to my innumerable friends and colleagues who, when I told them what I was working on, encouraged me, truthfully or not, by saying, "I'd love to read that book!"

Finally, writing a book becomes an all-consuming obsession over many years. However, that writing plays out in the context of a larger life, with its own demands and expectations. Without incredible support from my closest colleagues this book would not have emerged. I am deeply indebted to Leslie Price, my ever-ready and multidimensionally capable assistant, who over many years has created order from a chaotically vast array of requests and obligations. I thank the innumerable gracious colleagues and friends whose invitations I have declined, blaming the incessant demands of writing the book.

Most importantly, my husband, Tom, and our daughter, Elizabeth, have been my guiding stars and the loves of my life. They have thought with me, read with me, and perpetually centered me with their constant wisdom, insight, and love.

NOTES

PROLOGUE

ix **source: a revolutionary convergence:** P. Sharp, T. Jacks, and S. Hockfield, "Capitalizing on Convergence for Health Care," *Science* 352, no. 6293 (2016): 1522–23, http://doi.org/10.1126/science.aag2350; Phillip Sharp and Susan Hockfield, "Convergence: The Future of Health," *Science* 355, no. 6325 (2017): 589, http://doi.org/10.1126/science.aam8563.

x **Today's world population of around 7.6 billion:** United Nations Department of Economic and Social Affairs Population Division, "World Urbanization Prospects: The 2018 Revision," 2018, http://population.un.org/wup/DataQuery.

x **grappling with the consequences:** Chunwu Zhu et al.,"Carbon Dioxide (CO_2) Levels This Century Will Alter the Protein, Micronutrients, and Vitamin Content of Rice Grains with Potential Health Consequences for the Poorest Rice-Dependent Countries," *Science Advances* 4, no. 5 (2018): 1–8, http://doi.org/10.1126/sciadv.aaq1012.

x **Temperatures and sea levels are rising:** John A. Church and Neil J. White, "A 20th Century Acceleration in Global Sea-Level Rise," *Geophysical Research Letters* 33, no. 1 (2006): 94–97, http://doi.org/10.1029/2005GL024826; Benjamin D. Santer et al., "Tropospheric Warming over the Past Two Decades," *Scientific Reports* 7, no. 1 (2017): 3–8, http://doi.org/10.1038/s41598-017-02520-7.

x **the Reverend Thomas Robert Malthus:** Thomas Robert Malthus, "An Essay on the Principle of Population as It Affects the Future Improvement of Society," 1798.

x **England's population grew even more rapidly:** UK Census Online Project. Last modified May 29, 2015, http://www.freecen.org.uk.

xi **The technology-driven agricultural revolution:** Mark Overton, *Agricultural Revolution in England: The Transformation of the Agrarian Economy* (Cambridge: Cambridge University Press, 1996); Robert C. Allen, "Tracking the Agricultural Revolution in England," *Economic History Review* 52, no. 2 (1999): 209–35, http://doi.org/10.1111/1468-0289.00123.

xi **Biology and engineering are converging in previously unimaginable ways:** Susan Hockfield, "The Next Innovation Revolution," *Science* 323, no. 5918 (2009): 1147, http://doi.org/10.1126/science.1170834; Susan Hockfield, "A New Century's New Technologies," *Project Syndicate* (2015), http://www.project-syndicate.org/commentary/engineering-biotech-innovations-by-susan-hockfield-2015-01.

1: WHERE THE FUTURE COMES FROM

1 **At an early morning meeting of the MIT Corporation:** Marcella Bombardieri and Jenna Russell, "Female Leadership Signals Shift at MIT," *Boston Globe*, August 27, 2004; Arthur Jones, "Susan Hockfield Elected MIT's 16th President," *TechTalk* 49, no. 1 (2004); Katie Zezima, "M.I.T. Makes Yale Provost First Woman to Be Its Chief," *New York Times*, August 27, 2004, http://doi.org/10.13140/2.1.3945.0402.

2 **the dean of the School of Engineering reported:** Thomas Magnanti, in discussion with the author, fall 2004.

4 **to describe biological mechanisms:** H. F. Judson, *The Eighth Day of Creation: The Makers of the Revolution in Biology* (Plainview, NY: CSHL Press, 1996).

4 **James Watson, Francis Crick, Maurice Wilkins, and Rosalind Franklin:** Rosalind E. Franklin and R. G. Gosling, "Evidence for 2-Chain

Helix in Crystalline Structure of Sodium Deoxyribonucleate," *Nature* 172, no. 4369 (1953): 156–57; Rosalind E. Franklin and R. G. Gosling, "Molecular Configuration in Sodium Thymonucleate," *Nature* 171, no. 4356 (1953): 740–41; J. D. Watson and F. H. Crick, "Molecular Structure of Nucleic Acids: A Structure for Deoxyribose Nucleic Acid," *Nature* 171, no. 4356 (1953): 737–38; M. H. F. Wilkins, "Molecular Configuration of Nucleic Acids," *Science* 140, no. 3570 (1963): 941–50.

5 **a map of the human genome:** E. S. Lander et al., "Initial Sequencing and Analysis of the Human Genome," *Nature* 409, no. 6822 (2001): 860–921, http://doi.org/10.1038/35057062; J. C. Venter et al., "The Sequence of the Human Genome," *Science* 291, no. 5507 (2001): 1304–51, http://doi.org/10.1126/science.1058040.

6 **Watson also foresaw the possibility:** Cold Spring Harbor Symposia on Quantitative Biology, Molecular Neurobiology, XLVIII, C. S. H. Laboratory, 1983.

9 **In 1897, the great physicist J. J. Thomson:** Joseph John Thomson, "XL. Cathode Rays," *The London, Edinburgh, and Dublin Philosophical Magazine and Journal of Science* 44, no. 269 (1897): 293–316, http://doi.org/10.1080/14786449708621070.

9 **among them Marie and Pierre Curie, Wilhelm Roentgen, and Ernest Rutherford:** Ernest Rutherford, "LXXIX. The Scattering of α and β Particles by Matter and the Structure of the Atom," *Philosophical Magazine Series* 6, 21, no. 125 (1911): 669–88, http://doi.org/10.1080/14786440508637080; Otto Glasser, "W. C. Roentgen and the Discovery of the Roentgen Rays," *American Journal of Roentgenology* 165 (1995): 1033–40; R. F. Mould, "Marie and Pierre Curie and Radium: History, Mystery, and Discovery," *Medical Physics* 26, no. 9 (1999): 1766–72, http://doi.org/10.1118/1.598680.

9 **"We were awakening to":** National Academy of Sciences, Office of the Home Secretary, *Biographical Memoirs*, vol. 61 (Washington, DC: National Academy Press, 1992).

10 **he told the *Daily Princetonian:*** National Academy of Sciences, Office of the Home Secretary, *Biographical Memoirs*, vol. 61 (Washington, DC: National Academy Press, 1992).

10 **The Radiation Lab he helped create:** T. A. Saad, "The Story of the

M.I.T. Radiation Laboratory," *IEEE Aerospace and Electronic Systems Magazine* (October 1990): 46–51.

11 **Compton discussed this next convergence:** S. James Adelstein, "Robley Evans and What Physics Can Do for Medicine," *Cancer Biotherapy and Radiopharmaceuticals* 16, no. 3 (2001): 179–85, http://doi.org/10.1089/10849780152389375.

12 **incorporate radioactive labels into elements:** Angela N. H. Creager, "Phosphorus-32 in the Phage Group: Radioisotopes as Historical Tracers of Molecular Biology," *Studies in History and Philosophy of Biological and Biomedical Sciences* 40, no. 1 (2009): 29–42, http://doi.org/10.1016/j.shpsc.2008.12.005.Phosphorus-32.

12 **treating a set of patients with radioactive iodine:** S. Hertz, A. Roberts, and R. D. Evans, "Radioactive Iodine as an Indicator in the Study of Thyroid Physiology," *Proceedings of the Society for Experimental Biology and Medicine* 38 (1938): 510–13; S. Hertz and A. Roberts, "Radioactive Iodine in the Study of Thyroid Physiology: VII. The Use of Radioactive Iodine Therapy in Hyperthyroidism," *Journal of the American Medical Association* 131, no. 2 (1946): 81–86; Derek Bagley, "January 2016: Thyroid Month: The Saga of Radioiodine Therapy," *Endocrine News* (January 2016); Frederic H. Fahey, Frederick D. Grant, and James H. Thrall, "Saul Hertz, MD, and the Birth of Radionuclide Therapy," *EJNMMI Physics* 4, no. 1 (2017), http://doi.org/10.1186/s40658-017-0182-7.

12 **curriculum for Biological Engineering:** *MIT Reports to the President* 73, no. 1 (1937): 19–113; Karl T. Compton and John W. M. Bunker, "The Genesis of a Curriculum in Biological Engineering," *Scientific Monthly* 48, no. 1 (1939): 5–15.

12 **changed the name of MIT's Department of Biology:** *MIT Reports to the President* 80, no. 1 (1944): 8.

13 **He headed up American efforts:** National Academy of Sciences, Office of the Home Secretary, *Biographical Memoirs*, vol. 61 (Washington, DC: National Academy Press, 1992).

14 **Martin Polz, an environmental engineer:** Janelle R. Thompson et al., "Genotypic Diversity within a Natural Coastal Bacterioplankton Population," *Science* 307, no. 5713 (2005): 1311–13, http://doi.org/10.1126/

science.1106028; Dikla Man-Aharonovich et al., "Diversity of Active Marine Picoeukaryotes in the Eastern Mediterranean Sea Unveiled Using Photosystem-II psbA Transcripts," *ISME Journal* 4, no. 8 (2010): 1044–52, http://doi.org/10.1038/ismej.2010.25.

14 **Kristala Jones Prather, a chemical engineer:** Kristala Jones Prather et al., "Industrial Scale Production of Plasmid DNA for Vaccine and Gene Therapy: Plasmid Design, Production, and Purification," *Enzyme and Microbial Technology* 33, no. 7 (2003): 865–83, http://doi.org/10.1016/S0141-0229(03)00205-9; Kristala L. Jones Prather and Collin H. Martin, "De Novo Biosynthetic Pathways: Rational Design of Microbial Chemical Factories," *Current Opinion in Biotechnology* 19, no. 5 (2008): 468–74, http://doi.org/10.1016/j.copbio.2008.07.009; Micah J. Sheppard, Aditya M. Kunjapur, and Kristala L. J. Prather, "Modular and Selective Biosynthesis of Gasoline-Range Alkanes," *Metabolic Engineering* 33 (2016): 28–40, http://doi.org/10.1016/j.ymben.2015.10.010.

14 **Scott Manalis, a physicist turned biological engineer:** Thomas P. Burg et al., "Weighing of Biomolecules, Single Cells and Single Nanoparticles in Fluid," *Nature* 446, no. 7139 (2007): 1066–69, http://doi.org/10.1038/nature05741; Nathan Cermak et al., "High-Throughput Measurement of Single-Cell Growth Rates Using Serial Microfluidic Mass Sensor Arrays," *Nature Biotechnology* 34, no. 10 (2016): 1052–59, http://doi.org/10.1038/nbt.3666; Arif E. Cetin et al., "Determining Therapeutic Susceptibility in Multiple Myeloma by Single-Cell Mass Accumulation," *Nature Communications* 8, no. 1 (2017), http://doi.org/10.1038/s41467-017-01593-2.

14 **Robert Langer, regarded as the most prolific biological engineer:** Hannah Seligson, "Hatching Ideas, and Companies, by the Dozens at M.I.T.," *New York Times*, November 24, 2012, http://www.nytimes.com/2012/11/25/business/mit-lab-hatches-ideas-and-companies-by-the-dozens.html; Joel Brown, "MIT Scientist Robert Langer Talks about the Future of Research," *Boston Globe*, May 8, 2015, http://www.bostonglobe.com/magazine/2015/05/08/mit-scientist-robert-langer-talks-about-future-research/I0ggn93cxapR8omjcrM1hI/story.html.

2: CAN BIOLOGY BUILD A BETTER BATTERY?

19 **she proved the viability of her unconventional idea:** Sandra R. Whaley et al., "Selection of Peptides with Semiconductor Binding Specificity for Directed Nanocrystal Assembly," *Nature* 405, no. 6787 (2000): 665–68, http://doi.org/10.1038/35015043.

20 **in 2002, *Technology Review* named her:** "Innovators Under 35 2002: Angela Belcher," *MIT Technology Review*, 2002, http://www2.technologyreview.com/tr35/profile.aspx?trid=229.

20 **a MacArthur Foundation Genius Grant:** "MacArthur Fellows Program: Angela Belcher," 2004, http://www.macfound.org/fellows/727/.

20 ***Scientific American* named her Research Leader of the Year:** J. R. Minkel, "Scientific American 50: Research Leader of the Year," *Scientific American*, November 12, 2006, http://www.scientificamerican.com/article/scientific-american-50-re/.

22 **it arranges molecules of calcium carbonate:** A. M. Belcher et al., "Control of Crystal Phase Switching and Orientation by Soluble Mollusc-Shell Proteins," *Nature* 381, no. 56–58 (May 1996), http://doi.org/10.1038/381056a0.

22 **1/3,000 the strength of an abalone shell:** Bettye L. Smith et al., "Molecular Mechanistic Origin of the Toughness of Natural Adhesives, Fibers and Composites," *Nature* 399, no. 6738 (1999): 761–63, http://doi.org/10.1038/21607.

24 **"These translucent turquoise-colored crystals":** Stanislas Von Euw et al., "Biological Control of Aragonite Formation in Stony Corals," *Science* 356, no. 6341 (2017): 933–38, http://doi.org/10.1126/science.aam6371.

25 **Discoveries of bone and plant ashes:** F. Berna et al., "Microstratigraphic Evidence of in Situ Fire in the Acheulean Strata of Wonderwerk Cave, Northern Cape Province, South Africa," *Proceedings of the National Academy of Sciences* 109, no. 20 (2012): 1215–20, http://doi.org/10.1073/pnas.1117620109.

25 **the Neanderthals, used fire about 400,000 years ago:** W. Roebroeks and P. Villa, "On the Earliest Evidence for Habitual Use of Fire in Europe," *Proceedings of the National Academy of Sciences* 108, no. 13 (2011): 5209–14, http://doi.org/10.1073/pnas.1018116108.

25 **archeological evidence from Pech de l'Azé I:** Peter J. Heyes et al., "Selection and Use of Manganese Dioxide by Neanderthals," *Scientific Reports* 6 (2016), http://doi.org/10.1038/srep22159.

25 **photons from sunlight provide the source of energy:** Albert Einstein, "Über Einen Die Erzeugung Und Verwandlung Des Lichtes Betreffenden Heuristischen Gesichtspunkt," *Annalen der Physik (Leipzig)* 1905, http://doi.org/10.1002/pmic.201000799; A. B. Arons and M. B. Peppard, "Einstein's Proposal of the Photon Concept—A Translation of the *Annalen der Physik* Paper of 1905," *American Journal of Physics* 33, no. 5 (1964): 367–74.

26 **the primary source of energy in the United States:** Energy Information Administration Office of Energy Markets and End Use, *Annual Energy Review 2006*, 2007, http://doi.org/DOE/EIA-0384(2006).

26 **total US energy consumption:** U.S. Energy Information Administration, "U.S. Energy Facts Explained." Last modified May 19, 2017, http://www.eia.gov/energyexplained/?page=us_energy_home.

27 **After a long period of relative stability:** J. R. Petit et al., "Climate and Atmospheric History of the Past 420,000 Years from the Vostok Ice Core, Antartica," *Nature* 399, no. 6735 (1999): 429–36, https://doi.org/10.1038/20859; NASA Global Climate Change: Vital Signs of the Planet, "Graphic: The Relentless Rise of Carbon Dioxide." Last modified November 15, 2018, https://climate.nasa.gov/climate_resources/24/graphic-the-relentless-rise-of-carbon-dioxide/.

27 **close to 100 tons of plant material:** Jeffrey S. Dukes, "Burning Buried Sunshine: Human Consumption of Ancient Solar Energy," *Climatic Change* 61, no. 1–2 (2003): 31–44, http://doi.org/10.1023/A:1026391317686.

28 **Between 1950 and 1980, Chinese national policy:** Yuyu Chen et al., "Evidence on the Impact of Sustained Exposure to Air Pollution on Life Expectancy from China's Huai River Policy," *Proceedings of the National Academy of Sciences* 110, no. 32 (2013): 12936–41, http://doi.org/10.1073/pnas.1300018110/-/DCSupplemental.www.pnas.org/cgi/doi/10.1073/pnas.1300018110.

28 **The world's energy demand:** D. Larcher and J. M. Tarascon, "Towards Greener and More Sustainable Batteries for Electrical Energy Stor-

age," *Nature Chemistry* 7, no. 1 (2015): 19–29, http://doi.org/10.1038/nchem.2085.

28 **to grow from today's roughly 7.6 billion:** United Nations Department of Economic and Social Affairs Population Division, "World Urbanization Prospects: The 2018 Revision," 2018, http://population.un.org/wup/DataQuery.

28 **an American consumes, on average:** The World Bank Group, "Electric Power Consumption (kWh per capita)." Last modified 2014, http://data.worldbank.org/indicator/EG.USE.ELEC.KH.PC?locations=US.

28 **a Bangladeshi consumes, on average:** The World Bank Group, "Electric Power Consumption (kWh per capita)." Last modified 2014, http://data.worldbank.org/indicator/EG.USE.ELEC.KH.PC?locations=IN-PK-BD-LK-NP-AF.

29 **"Intermittency" has become the buzzword:** William A. Braff, Joshua M. Mueller, and Jessika E. Trancik, "Value of Storage Technologies for Wind and Solar Energy," *Nature Climate Change* 6, no. 10 (2016): 964–69, http://doi.org/10.1038/nclimate3045.

30 **The first battery dates to 1800:** P. A. Abetti, "The Letters of Alessandro Volta," *Electrical Engineering* 71, no. 9 (1952): 773–76, http://doi.org/10.1109/EE.1952.6437680.

32 **The first successful rechargeable battery:** P. Kurzweil, "Gaston Planté and His Invention of the Lead-Acid Battery—The Genesis of the First Practical Rechargeable Battery," *Journal of Power Sources* 195, no. 14 (2010): 4424–34, http://doi.org/10.1016/j.jpowsour.2009.12.126.

33 **vastly lighter and safer rechargeable batteries:** Bruno Scrosati and Jürgen Garche, "Lithium Batteries: Status, Prospects and Future," *Journal of Power Sources* 195, no. 9 (2010): 2419–30, http://doi.org/10.1016/j.jpowsour.2009.11.048; Akira Yoshino, "The Birth of the Lithium-Ion Battery," *Angewandte Chemie—International Edition* 51, no. 24 (2012): 5798–5800, http://doi.org/10.1002/anie.201105006.

33 **standard manufacturing processes for batteries:** Antti Väyrynen and Justin Salminen, "Lithium-Ion Battery Production," *Journal of Chemical Thermodynamics* 46 (2012): 80–85, http://doi.org/10.1016/j.jct.2011.09.005.

33 **the IVL Swedish Environmental Research Institute calculated:** Mia Romare and Lisbeth Dahllöf, "The Life Cycle Energy Consumption

and Greenhouse Gas Emissions from Lithium-Ion Batteries and Batteries for Light-Duty Vehicles," IVL Swedish Environmental Research Institute Report C 243, 2017.

33 **20 tons of carbon dioxide:** United States Environmental Protection Agency, "Greenhouse Gas Equivalencies Calculator." Last modified September 2017, http://www.epa.gov/energy/greenhouse-gas-equivalencies-calculator.

33 **2,250 gallons of gasoline are burned:** Tesla, "Model S: The Best Car." Last modified 2018, http://www.tesla.com/models.

34 **dozens of promising new technologies:** Sung Yoon Chung, Jason T. Bloking, and Yet Ming Chiang, "Electronically Conductive Phospho-Olivines as Lithium Storage Electrodes," *Nature Materials* 1, no. 2 (2002): 123–28, http://doi.org/10.1038/nmat732; Won Hee Ryu et al., "Heme Biomolecule as Redox Mediator and Oxygen Shuttle for Efficient Charging of Lithium-Oxygen Batteries," *Nature Communications* 7 (2016), http://doi.org/10.1038/ncomms12925.

34 **Yi Cui and his colleagues at Stanford:** Nian Liu et al., "A Pomegranate-Inspired Nanoscale Design for Large-Volume-Change Lithium Battery Anodes," *Nature Nanotechnology* 9, no. 3 (2014): 187–92, http://doi.org/10.1038/nnano.2014.6; Haotian Wang et al., "Direct and Continuous Strain Control of Catalysts with Tunable Battery Electrode Materials," *Science* 354, no. 6315 (2016): 1031–36.

34 **Ze Xiang Shen, at the Nanyang Technical University:** Dongliang Chao et al., "Array of Nanosheets Render Ultrafast and High-Capacity Na-Ion Storage by Tunable Pseudocapacitance," *Nature Communications* 7 (2016): 1–8, http://doi.org/10.1038/ncomms12122.

34 **little more than a protein capsule:** Nancy Trun and Janine Trempy, "Chapter 7: Bacteriophage," in *Fundamental Bacterial Genetics* (Hoboken, NJ: Wiley-Blackwell, 2003), 105–25, http://www.blackwellpublishing.com/trun/pdfs/Chapter7.pdf.

34 **dates back 300 million years:** Julien Thézé et al., "Paleozoic Origin of Insect Large dsDNA Viruses," *Proceedings of the National Academy of Sciences* 108, no. 38 (2011): 15931–35, https://doi.org/10.1073/pnas.1105580108.

36 **first reported the structure of DNA:** J. D. Watson and F. H. Crick,

"Molecular Structure of Nucleic Acids: A Structure for Deoxyribose Nucleic Acid," *Nature* 171, no. 4356 (1953): 737–38; H. F. Judson, *The Eighth Day of Creation: The Makers of the Revolution in Biology* (Plainview, NY: CSHL Press, 1996).

38 **Al Hershey and Margaret Chase famously used viruses:** A. D. Hershey and Martha Chase, "Independent Functions of Viral Protein and Nucleic Acid in Growth of Bacteriophage," *Journal of General Physiology* 36 (1952): 39–56; Angela N. H. Creager, "Phosphorus-32 in the Phage Group: Radioisotopes as Historical Tracers of Molecular Biology," *Studies in History and Philosophy of Biological and Biomedical Sciences* 40, no. 1 (2009): 29–42, http://doi.org/10.1016/j.shpsc.2008.12.005.Phosphorus-32.

42 **to produce virus-based battery electrodes:** Ki Tae Nam et al., "Genetically Driven Assembly of Nanorings Based on the M13 Virus," *Nano Letters* 4, no. 1 (2004): 23–27; Yu Huang et al., "Programmable Assembly of Nanoarchitectures Using Genetically Engineered Viruses," *Nano Letters* 5, no. 7 (2005): 1429–34, http://doi.org/10.1021/nl050795d; K. T. Nam et al., "Stamped Microbattery Electrodes Based on Self-Assembled M13 Viruses," *Proceedings of the National Academy of Sciences* 105, no. 45 (2008): 17227–31, http://doi.org/10.1073/pnas.0711620105; Dahyun Oh et al., "M13 Virus-Directed Synthesis of Nanostructured Metal Oxides for Lithium-Oxygen Batteries," *Nano Letters* 14, no. 8 (2014): 4837–45, http://doi.org/10.1021/nl502078m; Maryam Moradi et al., "Improving the Capacity of Sodium-Ion Battery Using a Virus-Templated Nanostructured Composite Cathode," *Nano Letters* 15, no. 5 (2015): 2917–21, http://doi.org/10.1021/nl504676v.

47 **the successful construction of a virus-enabled anode:** Ki Tae Nam et al., "Virus-Enabled Synthesis and Assembly of Nanowires for Lithium-Ion Battery Electrodes," *Science* 312, no. 5775 (2006): 885–88, http://doi.org/10.1126/science.1122716; Yun Jung Lee et al., "Biologically Activated Noble Metal Alloys at the Nanoscale: For Lithium Ion Battery Anodes," *Nano Letters* 10, no. 7 (2010): 2433–40, http://doi.org/10.1021/nl1005993.

47 **she did the same for a cathode:** Yun Jung Lee et al., "Fabricating Genetically Engineered High-Power Lithium-Ion Batteries Using Multiple Virus Genes," *Science* 324, no. 5930 (2009): 1051–55, http://

doi.org/10.1126/science.1171541; Dahyun Oh et al., "Biologically Enhanced Cathode Design for Improved Capacity and Cycle Life for Lithium-Oxygen Batteries," *Nature Communications* 4 (May 2013): 1–8, http://doi.org/10.1038/ncomms3756.

47 **President Barack Obama visited MIT:** David Chandler and Greg Frost, "Hockfield, Obama Urge Major Push in Clean Energy Research Funding," *MIT Tech Talk* 53, no. 20 (2009): 1–8.

47 **standard battery-manufacturing processes:** J. B. Dunn et al., "Material and Energy Flows in the Materials Production, Assembly, and End-of-Life Stages of the Automotive Lithium-Ion Battery Life Cycle," *Argonne National Laboratory Energy Systems Division* (2012), http://doi.org/10.1017/CBO9781107415324.004.

47 **150–200 kilograms of carbon dioxide:** Mia Romare and Lisbeth Dahllöf, "The Life Cycle Energy Consumption and Greenhouse Gas Emissions from Lithium-Ion Batteries and Batteries for Light-Duty Vehicles," IVL Swedish Environmental Research Institute Report C 243, 2017.

48 **world crude oil production more than doubled:** Energy Information Administration Office of Energy Markets and End Use, *Annual Energy Review 2006* (2007), http://doi.org/DOE/EIA-0384(2006).

3: WATER, WATER EVERYWHERE

49 **Peter Agre stumbled on a discovery:** Peter Agre, "The Aquaporin Water Channels," *Proceedings of the American Thoracic Society* 3 (2006): 5–13, http://doi.org/10.1513/pats.200510-109JH.

49 **the protein that caused Rh disease:** P. Agre and J. P. Cartron, "Molecular Biology of the Rh Antigens," *Blood* 78, no. 3 (1991): 551–63; Neil D. Avent and Marion E. Reid, "The Rh Blood Group System: A Review," *Blood* 95, no. 2 (2000): 375–87.

49 **He followed a classical strategy:** Peter Agre et al., "Purification and Partial Characterization of the Mr 30,000 Integral Membrane Protein Associated with the Erythrocyte Rh(D) Antigen," *Journal of Biological Chemistry* 262, no. 36 (1987): 17497–503; A. M. Saboori, B. L. Smith, and P. Agre, "Polymorphism in the Mr 32,000 Rh Protein Purified from Rh(D)-Positive and -Negative Erythrocytes," *Proceedings of the*

National Academy of Sciences of the United States of America 85, no. 11 (1988): 4042–45, http://doi.org/10.1073/pnas.85.11.4042.

50 **the 2003 Nobel Prize in Chemistry:** Peter Agre, "Aquaporin Water Channels: Nobel Lecture," 2003.

50 **more than 50 percent of our bodies:** H. H. Mitchell et al., "The Chemical Composition of the Adult Human Body and Its Bearing on the Biochemistry of Growth," *Journal of Biological Chemistry* 158 (1945): 625–37; ZiMian Wang et al., "Hydration of Fat-Free Body Mass: Review and Critique of a Classic Body-Composition Constant," *American Journal of Clinical Nutrition* 69 (1999): 833–841.

50 **some 300 million trillion (300×10^{18}) gallons:** USGS Water Science School, "How Much Water Is There on, in, and above the Earth?" Last modified December 2, 2016, http://water.usgs.gov/edu/earthhowmuch.html.

51 **more than 95 percent—is salty ocean water:** USGS Water Science School, "The World's Water." Last modified December 2, 2016, http://water.usgs.gov/edu/earthwherewater.html.

51 **More than 1 billion people lack access:** WWAP (United Nations World Water Assessment Programme), *The United Nations World Water Development Report 2015: Water for a Sustainable World* (Paris: UNESCO, 2015).

51 **by purifying the salt water and contaminated water:** Quirin Schiermeier, "Water: Purification with a Pinch of Salt," *Nature* 452, no. 7185 (2008): 260–61, http://doi.org/10.1038/452260a; Peter Gleick, "Why Don't We Get Our Drinking Water from the Ocean by Taking the Salt out of Seawater?," *Scientific American* Special Report: Confronting a World Freshwater Crisis (July 23, 2008); Ben Corry, "Designing Carbon Nanotube Membranes for Efficient Water Desalination," *Journal of Physical Chemistry B* 112, no. 5 (2008): 1427–34, http://doi.org/10.1021/jp709845u.

51 **Paintings from ancient Egypt:** George E. Symons, "Water Treatment through the Ages," *American Water Works Association* 98, no. 3 (2006): 87–97; Manish Kumar, Tyler Culp, and Yuexiao Shen, "Water Desalination: History, Advances, and Challenges," *The Bridge: Linking Engineering and Society* 46, no. 4 (2016): 21–29, http://doi.org/10.17226/24906.

51 **Aristotle described purification by distillation:** Aristotle, *Meteorology*, trans. E. W. Webster. 350 BCE. Available at http://classics.mit.edu/Aristotle/meteorology.1.i.html.

51 **water purification by distillation and filtration:** James E. Miller, "Review of Water Resources and Desalination Technologies," *SAND Report* (2003): 1–54, http://doi.org/SAND 2003-0800; T. M. Mayer, P. V. Brady, and R. T. Cygan, "Nanotechnology Applications to Desalination: A Report for the Joint Water Reuse and Desalination Task Force," *Sandia Report* (2011): 1–34; Muhammad Wakil Shahzad et al., "Energy-Water-Environment Nexus Underpinning Future Desalination Sustainability," *Desalination* 413 (2017): 52–64, http://doi.org/10.1016/j.desal.2017.03.009.

51 **Peter Agre's 1992 discovery**: Gregory M. Preston et al., "Appearance of Water Channels in Xenopus Oocytes Expressing Red Cell CHIP28 Protein," *Science* 256 (1992): 385–87, http://doi.org/http://science.sciencemag.org/content/256/5055/385.

51 **In 1988 Agre published a paper:** B. M. Denker et al., "Identification, Purification, and Partial Characterization of a Novel Mr 28,000 Integral Membrane Protein from Erythrocytes and Renal Tubules," *Journal of Biological Chemistry* 263, no. 30 (1988): 15634–42; M. P. De Vetten and P. Agre, "The Rh Polypeptide Is a Major Fatty Acid-Acylated Erythrocyte Membrane Protein," *Journal of Biological Chemistry* 263, no. 34 (1988): 18193–96.

52 **Parker, a clinical hematologist and oncologist:** Peter Agre, "Peter Agre—Biographical," in *Les Prix Nobel*, ed. Tore Frangsmyr (Stockholm: Nobel Foundation, 2004), http://www.nobelprize.org/nobel_prizes/chemistry/laureates/2003/agre-bio.html.

53 **35 trillion cells that make up our bodies:** Eva Bianconi et al., "An Estimation of the Number of Cells in the Human Body," *Annals of Human Biology* 40, no. 6 (2013): 463–71, http://doi.org/10.3109/03014460.2013.807878.

53 **Some researchers theorized that a unique channel protein:** Mario Parisi et al., "From Membrane Pores to Aquaporins: 50 Years Measuring Water Fluxes," *Journal of Biological Physics* 33, no. 5–6 (2007): 331–43, http://doi.org/10.1007/s10867-008-9064-5.

53 **Agre and Parker met up in 1991:** Peter Agre et al., "Aquaporin Water Channels—From Atomic Structure to Clinical Medicine," *Journal of Physiology* 542, no. 1 (2002): 3–16, http://doi.org/10.1113/jphysiol.2002.020818.

54 **as it diffuses across other filters:** G. Hummer, J. C. Rasaiah, and J. P. Noworyta, "Water Conduction through the Hydrophobic Channel of a Carbon Nanotube," *Nature* 414 (2001): 188–90.

54 **Agre and his colleagues identified specific DNA strands:** G. M. Preston and P. Agre, "Isolation of the cDNA for Erythrocyte Integral Membrane Protein of 28 Kilodaltons: Member of an Ancient Channel Family," *Proceedings of the National Academy of Sciences* 88, no. 24 (1991): 11110–14, http://doi.org/10.1073/pnas.88.24.11110.

55 **he injected the mystery protein's RNA:** Gregory M. Preston et al., "Appearance of Water Channels in Xenopus Oocytes Expressing Red Cell CHIP28 Protein," *Science* 256 (1992): 385–87, http://doi.org/http://science.sciencemag.org/content/256/5055/385.

55 **He named it aquaporin:** Peter Agre, Sei Sasaki, and Maarten J. Chrispeels, "Aquaporins: A Family of Water Channel Proteins," *Journal of Physiology* 265, no. 461 (1993): 92093; P. Agre et al., "Aquaporin CHIP: The Archetypal Molecular Water Channel," *American Journal of Physiology* 265 (1993): F463–76, http://doi.org/10.1085/jgp.79.5.791.

55 **a whole family of aquaporins:** Peter Agre, Dennis Brown, and Søren Nielsen, "Aquaporin Water Channels: Unanswered Questions and Unresolved Controversies," *Current Opinion in Cell Biology* 7, no. 4 (1995): 472–83, http://doi.org/10.1016/0955-0674(95)80003-4; Mario Borgnia et al., "Cellular and Molecular Biology of the Aquaporin Water Channels," *Annual Review of Biochemistry* 68 (1999): 425–58, http://doi.org/10.1177/154411130301400105.

56 **The structure of the aquaporin channel:** H. Sui et al., "Structural Basis of Water-Specific Transport through the AQP1 Water Channel," *Nature* 414, no. 6866 (2001): 872–78, http://doi.org/10.1038/414872a; Emad Tajkhorshid et al., "Control of the Selectivity of the Aquaporin Water Channel Family by Global Orientational Tuning," *Science* 296, no. 5567 (2002): 525–30, http://doi.org/10.1126/science.1067778; Dax Fu and Min Lu, "The Structural Basis of Water Permeation and Proton

Exclusion in Aquaporins," *Molecular Membrane Biology* 24, no. 5–6 (2007): 366–74, http://doi.org/10.1080/09687680701446965.

56 **Proteins are strings of beads:** B. Alberts et al., *Molecular Biology of the Cell*, 4th ed. (New York: Garland Science, 2002); The Shape and Structure of Proteins, http://www.ncbi.nlm.nih.gov/books/NBK26830/.

58 **the string winds up and loops around:** K. Murata et al., "Structural Determinants of Water Permeation through Aquaporin-1," *Nature* 407, no. 6804 (2000): 599–605; G. Ren et al., "Visualization of a Water-Selective Pore by Electron Crystallography in Vitreous Ice," *Proceedings of the National Academy of Sciences of the United States of America* 98, no. 4 (2001): 1398–1403, http://doi.org/10.1073/pnas.98.4.1398; Boaz Ilan et al., "The Mechanism of Proton Exclusion in Aquaporin Channels," *Proteins: Structure, Function and Genetics* 55, no. 2 (2004): 223–28, http://doi.org/10.1002/prot.20038.

58 **alternating negative and positive charges:** Bert L. De Groot et al., "The Mechanism of Proton Exclusion in the Aquaporin-1 Water Channel," *Journal of Molecular Biology* 333, no. 2 (2003): 279–93, http://doi.org/10.1016/j.jmb.2003.08.003; Fangqiang Zhu, Emad Tajkhorshid, and Klaus Schulten, "Theory and Simulation of Water Permeation in Aquaporin-1," *Biophysical Journal* 86, no. 1 (2004): 50–57, http://doi.org/10.1016/S0006-3495(04)74082-5; Xuesong Li et al., "Nature Gives the Best Solution for Desalination: Aquaporin-Based Hollow Fiber Composite Membrane with Superior Performance," *Journal of Membrane Science* 494 (2015): 68–77, http://doi.org/10.1016/j.memsci.2015.07.040.

59 **Almost every organism has them:** Tamir Gonen and Thomas Walz, "The Structure of Aquaporins," *Quarterly Reviews of Biophysics* 39, no. 4 (2006): 361–96, http://doi.org/10.1017/S0033583506004458.

60 **some members of the family also conduct other molecules:** D. Fu et al., "Structure of a Glycerol-Conducting Channel and the Basis for Its Selectivity," *Science* 290, no. 5491 (2000): 481–86, http://doi.org/10.1126/science.290.5491.481; B. L. De Groot and H. Grubmüller, "Water Permeation across Biological Membranes: Mechanism and Dynamics of Aquaporin-1 and GlpF," *Science* 294, no. 5550 (2001): 2353–57, http://doi.org/10.1126/science.1062459.

60 **water transport through plant roots:** Huayu Sun et al., "The Bamboo

Aquaporin Gene PeTIP4; 1–1 Confers Drought and Salinity Tolerance in Transgenic Arabidopsis," *Plant Cell Reports* 36, no. 4 (2017): 597–609, http://doi.org/10.1007/s00299-017-2106-3.

60 **water filtration in the kidney:** Landon S. King, David Kozono, and Peter Agre, "From Structure to Disease: The Evolving Tale of Aquaporin Biology," *Nature Reviews Molecular Cell Biology* 5 (2004): 687–98.

61 **home to more than 9.5 billion people:** United Nations Department of Economic and Social Affairs Population Division, "World Urbanization Prospects: The 2018 Revision," 2018, http://population.un.org/wup/DataQuery.

61 **I flew to Denmark to visit Hélix-Nielsen:** Hélix-Nielsen, Claus and Peter Holme Jensen in discussion with the author, September 2017.

62 **the promise of matching the cost and efficiency of current systems:** C. Y. Tang et al., "Desalination by Biomimetic Aquaporin Membranes: Review of Status and Prospects," *Desalination* 308 (2013): 34–40, http://doi.org/10.1016/j.desal.2012.07.007; Mariusz Grzelakowski et al., "A Framework for Accurate Evaluation of the Promise of Aquaporin Based Biomimetic Membranes," *Journal of Membrane Science* 479 (2015): 223–31, http://doi.org/10.1016/j.memsci.2015.01.023.

63 **A red blood cell measures less than 10 micrometers:** M. Dao, C. T. Lim, and S. Suresh, "Mechanics of the Human Red Blood Cell Deformed by Optical Tweezers," *Journal of the Mechanics and Physics of Solids* 51, no. 11–12 (2003): 2259–80, http://doi.org/10.1016/j.jmps.2003.09.019.

63 **as thick as a dime:** United States Mint, "Coin Specifications." Last modified April 5, 2018, http://www.usmint.gov/learn/coin-and-medal-programs/coin-specifications.

63 **blockbuster drugs of the twentieth century:** Arne E. Brändström and Bo R. Lamm, Processes for the preparation of omeprazole and intermediates therefore, issued 1985, http://doi.org/US005485919A; Bruce D. Roth, Trans-6-2-(3- OR 4-Carboxamide-substituted pyrrol-1-yl)alkyl-4-hydroxypyran-2-one inhibitors of cholesterol synthesis, issued 1987, http://doi.org/10.1016/j.(73); W. Sneader, "The Discovery of Aspirin" *Pharmaceutical Journal* 259, no. 6964 (1997): 614–17, http://doi.org/10.1136/bmj.321.7276.1591; Kay Brune, B. Renner, and G. Tiegs,

"Acetaminophen/Paracetamol: A History of Errors, Failures and False Decisions," *European Journal of Pain (United Kingdom)* 19, no. 7 (2015): 953–65, http://doi.org/10.1002/ejp.621.

64 **protein-based drugs such as insulin and growth hormone:** D. V. Goeddel et al., "Expression in Escherichia Coli of Chemically Synthesized Genes for Human Insulin," *Proceedings of the National Academy of Sciences of the United States of America* 76, no. 1 (1979): 106–10, http://doi.org/10.1073/pnas.76.1.106; Henrik Dalbøge et al., "A Novel Enzymatic Method for Production of Authentic HGH from an Escherichia Coli Produced HGH-Precursor," *Nature Biotechnology* 5 (1987): 161–64; Mohamed N. Baeshen, "Production of Biopharmaceuticals in *E. Coli*: Current Scenario and Future Perspectives," *Journal of Microbiology and Biotechnology* 25, no. 7 (2014): 1–24, http://doi.org/10.4014/jmb.1405.05052.

64 **naturally makes its own aquaporin:** Giuseppe Calamita et al., "Molecular Cloning and Characterization of AqpZ, a Water Channel from Escherichia Coli," *Journal of Biological Chemistry* 270, no. 49 (1995): 29063–66, http://doi.org/10.1074/jbc.270.49.29063.

65 **the cell machinery can repair or replace defective proteins:** F. Ulrich Hartl, Andreas Bracher, and Manajit Hayer-Hartl, "Molecular Chaperones in Protein Folding and Proteostasis," *Nature* 475, no. 7356 (2011): 324–32, http://doi.org/10.1038/nature10317.

65 **stable enough to live for about four months:** David Shemin and D. Rittenberg, "The Life Span of the Human Red Blood Cell," *Journal of Biological Chemistry* 166 (1946): 627–36.

65 **skin cells that live less than a month:** Gerald D. Weinstein and Eugene J. van Scott, "Autoradiographic Analysis of Turnover Times of Normal and Psoriatic Epidermis," *Journal of Investigative Dermatology* 45, no. 4 (1965): 257–62, http://doi.org/10.1038/jid.1965.126.

65 **cells lining the digestive system:** H. J. Li et al., "Basic Helix-Loop-Helix Transcription Factors and Enteroendocrine Cell Differentiation," *Diabetes, Obesity and Metabolism* 13, Suppl 1, no. 2 (2011): 5–12, http://doi.org/10.1111/j.1463-1326.2011.01438.x.

66 **it can survive the extreme temperatures:** Xuesong Li et al., "Preparation of High Performance Nanofiltration (NF) Membranes Incor-

porated with Aquaporin Z," *Journal of Membrane Science* 450 (2014): 181–88, http://doi.org/10.1016/j.memsci.2013.09.007.

66 **chemical conditions required to purify it:** Saren Qi et al., "Aquaporin-Based Biomimetic Reverse Osmosis Membranes: Stability and Long Term Performance," *Journal of Membrane Science* 508 (2016): 94–103, http://doi.org/10.1016/j.memsci.2016.02.013.

66 **re-embed the isolated aquaporin proteins in membranes:** Yan Zhao et al., "Synthesis of Robust and High-Performance Aquaporin-Based Biomimetic Membranes by Interfacial Polymerization-Membrane Preparation and RO Performance Characterization," *Journal of Membrane Science* 423–424 (2012): 422–28, http://doi.org/10.1016/j.memsci.2012.08.039; Honglei Wang, Tai Shung Chung, and Yen Wah Tong, "Study on Water Transport through a Mechanically Robust Aquaporin Z Biomimetic Membrane," *Journal of Membrane Science* 445 (2013): 47–52, http://doi.org/10.1016/j.memsci.2013.05.057.

69 **a vesicle sheet that was cheaper and stronger:** Yang Zhao et al., "Effects of Proteoliposome Composition and Draw Solution Types on Separation Performance of Aquaporin-Based Proteoliposomes: Implications for Seawater Desalination Using Aquaporin-Based Biomimetic Membranes," *Environmental Science and Technology* 47, no. 3 (2013): 1496–1503, http://doi.org/10.1021/es304306t.

69 **A single, flat cell membrane:** Honglei Wang, Tai Shung Chung, and Yen Wah Tong, "Study on Water Transport through a Mechanically Robust Aquaporin Z Biomimetic Membrane," *Journal of Membrane Science* 445 (2013): 47–52, http://doi.org/10.1016/j.memsci.2013.05.057; Chuyang Tang et al., "Biomimetic Aquaporin Membranes Coming of Age," *Desalination* 368 (2015): 89–105, http://doi.org/10.1016/j.desal.2015.04.026.

69 **a method to layer the aquaporin vesicle sheet:** Zhaolong Hu, James C. S. Ho, and Madhavan Nallani, "Synthetic (Polymer) Biology (Membrane): Functionalization of Polymer Scaffolds for Membrane Protein," *Current Opinion in Biotechnology* 46 (2017): 51–56, http://doi.org/10.1016/j.copbio.2016.10.012; Marta Espina Palanco et al., "Tuning Biomimetic Membrane Barrier Properties by Hydrocarbon, Cholesterol and Polymeric Additives," *Bioinspiration and Biomimetics* 13, no. 1 (2017): 1–11, http://doi.org/10.1088/1748-3190/aa92be.

69 **Danish astronauts used Aquaporin A/S membranes:** "Aquaporin Inside Membranes Undergo Second Round of Test in Space," *Membrane Technology* (February 2017): 5–6, http://doi.org/10.1016/S0958-2118(17)30032-0; "Aquaporin Inside Membrane Testing in Space (AquaMembrane)," NASA International Space Station Research and Technology. Last modified October 4, 2017, http://www.nasa.gov/mission_pages/station/research/experiments/2156.html.

71 **70 percent of the planet's freshwater:** WWAP (United Nations World Water Assessment Programme), *The United Nations World Water Development Report 2015: Water for a Sustainable World* (Paris: UNESCO, 2015).

4: CANCER-FIGHTING NANOPARTICLES

73 **the United States launched the War on Cancer:** National Cancer Institute, "National Cancer Act of 1971." Last modified February 16, 2016, http://www.cancer.gov/about-nci/legislative/history/national-cancer-act-1971; Eliot Marshall, "Cancer Research and the $90 Billion Metaphor," *Science* 331, no. 6024 (2011): 1540–41, http://doi.org/10.1126/science.331.6024.1540-a.

73 **600,000 Americans and more than 8 million people worldwide:** Rebecca L. Siegel, Kimberly D. Miller, and Ahmedin Jemal, "Cancer Statistics, 2018." *CA: A Cancer Journal for Clinicians* 68, no. 1 (2018): 7–30, http://doi.org/10.3322/caac.21442; National Cancer Institute, "Cancer Statistics." Last modified April 27, 2018, http://www.cancer.gov/about-cancer/understanding/statistics.

74 **identified a gene from the Rous Sarcoma Virus:** P. Rous, "A Transmissible Avian Neoplasm. (Sarcoma of the Common Fowl)," *Journal of Experimental Medicine* 12 (1910): 696–705, http://doi.org/10.1084/jem.12.5.696; P. Rous, "A Sarcoma of the Fowl Transmissible by an Agent Separable from the Tumor Cells," *Journal of Experimental Medicine* 13 (1911): 397–411, http://doi.org/10.1097/00000441-191108000-00079; Robin A. Weiss and Peter K. Vogt, "100 Years of Rous Sarcoma Virus," *Journal of Experimental Medicine* 208, no. 12 (2011): 2351–55, http://doi.org/10.1084/jem.20112160.

74 **cancer can also arise from intrinsic sources:** Marco A. Pierotti, Gabriella Sozzi, and Carlo M. Croce, "Discovery and Identification of Oncogenes," in *Holland-Frei Cancer Medicine*, ed. D. W. Kufe, R. E. Pollock, and R. R. Weichselbaum, 6th ed. (Hamilton: BC Decker, 2003); Peter K. Vogt, "Retroviral Oncogenes: A Historical Primer," *Nature Reviews Cancer* 12, no. 9 (2012): 639–48, http://doi.org/10.1038/nrc3320.Retroviral; Klaus Bister, "Discovery of Oncogenes: The Advent of Molecular Cancer Research," *Proceedings of the National Academy of Sciences* 112, no. 50 (2015): 15259–60, http://doi.org/10.1073/pnas.1521145112.

75 **Gleevec blocks the action of that protein:** Andrew Z. Wang, Robert Langer, and Omid C. Farokhzad, "Nanoparticle Delivery of Cancer Drugs," *Annual Review of Medicine* 63, no. 1 (2012): 185–98, http://doi.org/10.1146/annurev-med-040210-162544.

75 **Gleevec has increased the survival rate for patients:** Andreas Hochhaus et al., "Long-Term Outcomes of Imatinib Treatment for Chronic Myeloid Leukemia," *New England Journal of Medicine* 376, no. 10 (2017): 917–27, http://doi.org/10.1056/NEJMoa1609324.

76 **more than a third of all cancers today are preventable:** World Health Organization, "Cancer Prevention." Last modified 2018, http://www.who.int/cancer/prevention/en/

76 **mammography and colonoscopy screenings:** Sidney J. Winawer et al., "Colorectal Cancer Screening: Clinical Guidelines and Rationale: The Adenoma-Carcinoma Sequence," *Gastroenterology* 112 (1997): 594–642, http://doi.org/10.1053/GAST.1997.V112.AGAST970594; M. G. Marmot et al., "The Benefits and Harms of Breast Cancer Screening: An Independent Review," *British Journal of Cancer* 108, no. 11 (2013): 2205–40, http://doi.org/10.1038/bjc.2013.177.

76 **the five-year survival rate for breast cancer:** A. M. Noone et al., eds., SEER Cancer Statistics Review, 1975–2015, National Cancer Institute. Bethesda, MD, http://seer.cancer.gov/csr/1975_2015/, based on November 2017 SEER data submission, posted to the SEER website, April 2018; "Cancer Stat Facts: Female Breast Cancer," National Cancer Institute Surveillance, Epidemiology, and End Results Program. Last modified 2015, http://seer.cancer.gov/statfacts/html/breast.html.

77 **and for colon cancer patients:** "Cancer Stat Facts: Colorectal Cancer," National Cancer Institute Surveillance, Epidemiology, and End Results Program. Last modified 2015, http://seer.cancer.gov/statfacts/html/colorect.html.

77 **current standard imaging techniques:** John V. Frangioni, "New Technologies for Human Cancer Imaging," *Journal of Clinical Oncology* 26, no. 24 (2008): 4012–21, http://doi.org/10.1200/JCO.2007.14.3065.

77 **Blood-based detection tests face the same challenges:** N. Lynn Henry and Daniel F. Hayes, "Cancer Biomarkers," *Molecular Oncology* 6, no. 2 (2012): 140–46, http://doi.org/10.1016/j.molonc.2012.01.010.

78 **Bhatia has devised a urine-based test:** Gabriel A. Kwong et al., "Mass-Encoded Synthetic Biomarkers for Multiplexed Urinary Monitoring of Disease," *Nature Biotechnology* 31, no. 1 (2013): 63–70, http://doi.org/10.1038/nbt.2464.

78 **cell masses are as much as twenty times smaller**: Ester J. Kwon, Jaideep S. Dudani, and Sangeeta N. Bhatia, "Ultrasensitive Tumour-Penetrating Nanosensors of Protease Activity," *Nature Biomedical Engineering* 1, no. 4 (2017), http://doi.org/10.1038/s41551-017-0054.

79 **In her graduate work she used tools**: S. N. Bhatia et al., "Selective Adhesion of Hepatocytes on Patterned Surfaces," *Annals of the New York Academy of Sciences* 745 (1994): 187–209, http://www3.interscience.wiley.com/journal/119271052/abstract%5Cnpapers://e7896fb4-5763-415a-bb1b-292b8a8ba273/Paper/p904.

79 **a nanotechnologist with a knack for solving biomedical problems:** Austin M. Derfus, Warren C. W. Chan, and Sangeeta N. Bhatia, "Probing the Cytotoxicity of Semiconductor Quantum Dots," *Nano Letters* 4, no. 1 (2004): 11–18, http://doi.org/10.1021/nl0347334.

79 **Nanoparticles are simply very tiny pieces of matter:** Jörg Kreuter, "Nanoparticles—A Historical Perspective," *International Journal of Pharmaceutics* 331, no. 1 (2007): 1–10, http://doi.org/10.1016/j.ijpharm.2006.10.021.

79 **iron oxide is a very useful material for magnetic resonance imaging:** Saeid Zanganeh et al., "The Evolution of Iron Oxide Nanoparticles for Use in Biomedical MRI Applications," *SM Journal Clinical and Medical Imaging* 2, no. 1 (2016): 1–11.

80 **Gold nanoparticles, for example, don't appear gold:** Florian J. Heiligtag and Markus Niederberger, "The Fascinating World of Nanoparticle Research," *Materials Today* 16, no. 7–8 (2013): 262–71, http://doi.org/10.1016/j.mattod.2013.07.004.

80 **Roman artisans unwittingly created gold nanoparticles:** Ian Freestone et al., "The Lycurgus Cup—A Roman Nanotechnology," *Gold Bulletin* 40, no. 4 (2007): 270–77.

80 **Cadmium selenide naturally forms large black crystals:** Debasis Bera et al., "Quantum Dots and Their Multimodal Applications: A Review," *Materials* 3, no. 4 (2010): 2260–2345, http://doi.org/10.3390/ma3042260.

80 **we put silver nanoparticles into toothpaste:** Jonas Junevi, Juozas Žilinskas, and Darius Gleiznys, "Antimicrobial Activity of Silver and Gold in Toothpastes: A Comparative Analysis," *Stomatologija, Baltic Dental and Maxillofacial Journal* 17, no. 1 (2015): 9–12.

80 **We put titanium oxide and zinc oxide nanoparticles into sunblock:** Florian J. Heiligtag and Markus Niederberger, "The Fascinating World of Nanoparticle Research," *Materials Today* 16, no. 7–8 (2013): 262–71, http://doi.org/10.1016/j.mattod.2013.07.004.

81 **Sailor, a chemist and materials scientist:** Geoffrey Von Maltzahn et al., "Nanoparticle Self-Assembly Gated by Logical Proteolytic Triggers," *Journal of the American Chemical Society* 129, no. 19 (2007): 6064–65, http://doi.org/10.1021/ja070461l; Ji Ho Park et al., "Magnetic Iron Oxide Nanoworms for Tumor Targeting and Imaging," *Advanced Materials* 20, no. 9 (2008): 1630–35, http://doi.org/10.1002/adma.200800004.

81 **Ruoslahti, now based at the Stanford-Burnham Institute:** M. E. Akerman et al., "Nanocrystal Targeting in Vivo," *Proceedings of the National Academy of Sciences* 99, no. 20 (2002): 12617–21, http://doi.org/10.1073/pnas.152463399; Kazuki N. Sugahara et al., "Co-Administration of a Tumor-Penetrating Peptide Enhances the Efficacy of Cancer Drugs," *Science* 328, no. 5981 (2010): 1031–35, http://doi.org/10.1126/science.1183057; Ester J. Kwon et al., "Porous Silicon Nanoparticle Delivery of Tandem Peptide Anti-Infectives for the Treatment of Pseudomonas Aeruginosa Lung Infections,"

Advanced Materials 29, no. 35 (2017): 1–9, http://doi.org/10.1002/adma.201701527.

82 **they had to create clusters:** Todd J. Harris et al., "Proteolytic Actuation of Nanoparticle Self-Assembly," *Angewandte Chemie—International Edition* 45, no. 19 (2006): 3161–65, http://doi.org/10.1002/anie.200600259.

83 **they made two sets of nanoparticles:** Todd J. Harris et al., "Protease-Triggered Unveiling of Bioactive Nanoparticles," *Small* 4, no. 9 (2008): 1307–12, http://doi.org/10.1002/smll.200701319.

84 **They come in thousands of varieties:** A. Bairoch, "The ENZYME Database in 2000," *Nucleic Acids Research* 28, no. 1 (2000): 304–5, http://doi.org/10.1093/nar/28.1.304.

85 **Some enzymes can cut as many as a thousand:** Arren Bar-Even et al., "The Moderately Efficient Enzyme: Evolutionary and Physicochemical Trends Shaping Enzyme Parameters," *Biochemistry* 50, no. 21 (2011): 4402–10, http://doi.org/10.1021/bi2002289.

86 **In 2006, she and her team reported:** Dmitri Simberg et al., "Biomimetic Amplification of Nanoparticle Homing to Tumors," *Proceedings of the National Academy of Sciences* 104, no. 3 (2007): 932–36, http://doi.org/10.1073/pnas.0610298104.

86 **in 2009, they demonstrated tissue-specific delivery:** Todd J. Harris et al., "Tissue-Specific Gene Delivery via Nanoparticle Coating," *Biomaterials* 31, no. 5 (2010): 998–1006, http://doi.org/10.1016/j.biomaterials.2009.10.012.

87 **the natural tissue and molecular barriers:** Elvin Blanco, Haifa Shen, and Mauro Ferrari, "Principles of Nanoparticle Design for Overcoming Biological Barriers to Drug Delivery," *Nature Biotechnology* 33, no. 9 (2015): 941–51, http://doi.org/10.1038/nbt.3330.

88 **an unexpected fluorescent signal:** Andrew D. Warren et al., "Disease Detection by Ultrasensitive Quantification of Microdosed Synthetic Urinary Biomarkers," *Journal of the American Chemical Society* 136 (2014): 13709–14, http://doi.org/10.1021/ja505676h; Simone Schuerle et al., "Magnetically Actuated Protease Sensors for in Vivo Tumor Profiling," *Nano Letters* 16, no. 10 (2016): 6303–10, http://doi.org/10.1021/acs.nanolett.6b02670.

89 **very early cancer diagnosis:** Jaideep S. Dudani et al., "Classification of Prostate Cancer Using a Protease Activity Nanosensor Library," *Proceedings of the National Academy of Sciences* 115, no. 36 (2018): 8954–59, http://doi.org/10.1073/pnas.1805337115.

90 **detect tumors smaller than five millimeters:** Gabriel A. Kwong et al., "Mathematical Framework for Activity-Based Cancer Biomarkers," *Proceedings of the National Academy of Sciences* 112, no. 41 (2015): 12627–32, http://doi.org/10.1073/pnas.1506925112.

90 **tumors can take many years to grow:** Sharon S. Hori and Sanjiv S. Gambhir, "Mathematical Model Identifies Blood Biomarker-Based Early Cancer Detection Strategies and Limitations," *Science Translational Medicine* 3, no. 109 (2011), http://doi.org/10.1126/scitranslmed.3003110.Mathematical.

90 **a market-ready urine test:** A. D. Warren et al., "Point-of-Care Diagnostics for Noncommunicable Diseases Using Synthetic Urinary Biomarkers and Paper Microfluidics," *Proceedings of the National Academy of Sciences* 111, no. 10 (2014): 3671–76, http://doi.org/10.1073/pnas.1314651111.

90 **nanoparticles that release their drug payloads slowly:** Zhou J. Deng et al., "Layer-by-Layer Nanoparticles for Systemic Codelivery of an Anticancer Drug and SiRNA for Potential Triple-Negative Breast Cancer Treatment," *ACS Nano* 7, no. 11 (2013): 9571–84, http://doi.org/10.1021/nn4047925; Erkki Ruoslahti, Sangeeta N. Bhatia, and Michael J. Sailor, "Targeting of Drugs and Nanoparticles to Tumors," *Journal of Cell Biology* 188, no. 6 (2010): 759–68, http://doi.org/10.1083/jcb.200910104; Zvi Yaari et al., "Theranostic Barcoded Nanoparticles for Personalized Cancer Medicine," *Nature Communications* 7 (2016), http://doi.org/10.1038/ncomms13325; Rong Tong et al., "Photoswitchable Nanoparticles for Triggered Tissue Penetration and Drug Delivery," *Journal of the American Chemical Society* 134, no. 21 (2012): 8848–55, http://doi.org/10.1021/ja211888a; Dan Peer et al., "Nanocarriers as an Emerging Platform for Cancer Therapy," *Nature Nanotechnology* 2, no. 12 (2007): 751–60, http://doi.org/10.1038/nnano.2007.387.

90 **enhance the power of medical ultrasound and MRI:** Melodi Javid Whitley et al., "A Mouse-Human Phase 1 Co-Clinical Trial of

a Protease-Activated Fluorescent Probe for Imaging Cancer," *Science Translational Medicine* 8, no. 320 (2016): 4–6, http://doi.org/10.1126/scitranslmed.aad0293.

5: AMPLIFYING THE BRAIN

91 **He was scaling the cliffs:** Jim Ewing in discussion with the author, May 2018.

92 **lost both of his lower legs in a mountaineering accident:** Eric Moskowitz, "The Prosthetic of the Future," *Boston Globe*, November 21, 2016, http://www.bostonglobe.com/metro/2016/11/21/the-prosthetic-future/Ld6C2rxZL4uiotc96kNyPO/story.html.

92 **he began to make prosthetic legs:** Hugh Herr in discussions with the author, 2006–2018.

93 **The nerve cells that directly activate muscles:** "Motor Neurons," *PubMed Health Glossary*, http://www.ncbi.nlm.nih.gov/pubmedhealth/PMHT0024358/; Andrew B. Schwartz, "Movement: How the Brain Communicates with the World," *Cell* 164, no. 6 (2016): 1122–35, http://doi.org/10.1016/j.cell.2016.02.038.

95 **Herr studied the biology of walking:** Hugh M. Herr and Alena M. Grabowski, "Bionic Ankle—Foot Prosthesis Normalizes Walking Gait for Persons with Leg Amputation," *Proceedings of the Royal Society B* 279 (2012): 457–64, http://doi.org/10.1098/rspb.2011.1194.

95 **measured the energy a person uses in walking:** Samuel K. Au, Jeff Weber, and Hugh Herr, "Powered Ankle—Foot Prosthesis Improves Walking Metabolic Economy," *IEEE Transactions on Robotics* 25, no. 1 (2009); Luke M. Mooney, Elliott J. Rouse, and Hugh M. Herr, "Autonomous Exoskeleton Reduces Metabolic Cost of Human Walking," *Journal of NeuroEngineering and Rehabilitation* 11, no. 1 (2014): 1–5, http://doi.org/10.1186/1743-0003-11-151.

96 **I paid a visit to one of the companies:** Hildur Einarsdottir, Kim De Roy, and Magnus Oddsson in discussion with the author, October 2017.

104 **Restoring mobility after brain injury:** Beata Jarosiewicz et al., "Virtual Typing by People with Tetraplegia Using a Self-Calibrating Intracortical Brain-Computer Interface," *Science Translational Medicine* 7,

no. 313 (2015): 1–11; B. Wodlinger et al., "Ten-Dimensional Anthropomorphic Arm Control in a Human Brain–Machine Interface: Difficulties, Solutions, and Limitations," *Journal of Neural Engineering* 12, no. 1 (2015), http://doi.org/10.1088/1741-2560/12/1/016011; S. R. Soekadar et al., "Hybrid EEG/EOG-Based Brain/Neural Hand Exoskeleton Restores Fully Independent Daily Living Activities after Quadriplegia," *Science Robotics* 1 (2016): 1–8.

104 **I traveled to Geneva, Switzerland:** John Donoghue in discussion with the author, September 2017; Jens Clausen et al., "Help, Hope, and Hype: Ethical Dimensions of Neuroprosthetics," *Science* 356, no. 6345 (2017): 1338–39.

104 **how the brain orchestrates physical movements:** D. Purves et al., eds., "The Primary Motor Cortex: Upper Motor Neurons That Initiate Complex Voluntary Movements," in *Neuroscience*, 2nd ed. (Sunderland, MA: Sinauer Associates, 2001), http://www.ncbi.nlm.nih.gov/books/NBK10962/.

105 **each point in the PMC corresponds:** John P. Donoghue and Steven P. Wise, "The Motor Cortex of the Rat: Cytoarchitecture and Microstimulation Mapping," *Journal of Comparative Neurology* 212 (1982): 76–88; Shy Shoham et al., "Statistical Encoding Model for a Primary Motor Cortical Brain-Machine Interface," *IEEE Transactions on Biomedical Engineering* 52, no. 7 (2005): 1312–22; T. Aflalo et al., "Decoding Motor Imagery from the Posterior Parietal Cortex of a Tetraplegic Human," *Science* 348, no. 6237 (2015): 906–10, http://doi.org/10.7910/DVN/GJDUTV.

107 **an awareness of pain, heat, or cold or of the position:** Sharlene N. Flesher et al., "Intracortical Microstimulation of Human Somatosensory Cortex," *Science Translational Medicine* 8 (2016): 1–11; Emily L. Graczyk et al., "The Neural Basis of Perceived Intensity in Natural and Artificial Touch," *Science Translational Medicine* 142 (2016): 1–11; Luke E. Osborn et al., "Prosthesis with Neuromorphic Multilayered E-Dermis Perceives Touch and Pain," *Science Robotics* 3 (2018): 1–11, http://doi.org/10.1126/scirobotics.aat3818.

108 **they developed an intracortical brain-computer interface:** John P. Donoghue, "Connecting Cortex to Machines: Recent Advances in Brain Interfaces," *Nature Neuroscience* 5, no. 11 (2002): 1085–88, http://

doi.org/10.1038/nn947; Mijail D. Serruya et al., "Instant Neural Control of a Movement Signal," *Nature* 416, no. 6877 (2002): 141–42, http://doi.org/10.1038/416141a; Vicki Brower, "When Mind Meets Machine," *EMBO Reports* 6, no. 2 (2005): 108–10.

108 **showcased the potential of their iBCI:** Leigh R. Hochberg et al., "Neuronal Ensemble Control of Prosthetic Devices by a Human with Tetraplegia," *Nature* 442 (July 2006), http://doi.org/10.1038/nature04970.

109 **they reported that a woman named Cathy:** Leigh R. Hochberg et al., "Reach and Grasp by People with Tetraplegia Using a Neurally Controlled Robotic Arm," *Nature* 485, no. 7398 (2012: 372–75, http://doi.org/10.1038/nature11076; Andrew Jackson, "Neuroscience: Brain-Controlled Robot Grabs Attention," *Nature* 485, no. 7398 (2012): 317–18, http://doi.org/10.1038/485317a.

110 **The team captured Cathy on video:** "Paralyzed Woman Moves Robot with Her Mind," *Nature Video*. Last modified May 16, 2012, http://www.youtube.com/watch?v=ogBX18maUiM.

110 **In 2017, Donoghue and his team:** A. Bolu Ajiboye et al., "Restoration of Reaching and Grasping in a Person with Tetraplegia through Brain-Controlled Muscle Stimulation: A Proof-of-Concept Demonstration," *Lancet* 389 (2017): 1821–30, http://doi.org/10.1016/S0140-6736(17)30601-3; Clive Cookson, "Paralysed Man Regains Arm Movement Using Power of Thought," *Financial Times*, March 28, 2017, http://www.ft.com/content/1460d6e6-10c0-11e7-b030-768954394623; "Using Thought to Control Machines: Brain-Computer Interfaces May Change What It Means to Be Human," *The Economist*, January 4, 2018, http://www.economist.com/leaders/2018/01/04/using-thought-to-control-machines.

111 **The chip measures about 4 millimeters:** Leigh R. Hochberg et al., "Neuronal Ensemble Control of Prosthetic Devices by a Human with Tetraplegia," *Nature* 442 (July 2006), http://doi.org/10.1038/nature04970.

111 **Since the late 1960s, when brain-computer interfaces:** Karl Frank, "Some Approaches to the Technical Problem of Chronic Excitation of Peripheral Nerve" (speech), April 1968, Centennial Celebration of the American Otological Society.

112 **Dr. Leigh Hochberg, envisions:** Leigh Hochberg in discussion with the author, December 2017; Bob Tedeschi, "When Might Patients Use Their Brains to Restore Movement? 'We All Want the Answer to Be Now,'" *STAT,* June 6, 2017, http://www.statnews.com/2017/06/02/braingate-movement-paralysis/.

112 **restore normal life to those disabled:** Chethan Pandarinath et al., "High Performance Communication by People with Paralysis Using an Intracortical Brain-Computer Interface," *ELIFE* 6 (2017): 1–27, http://doi.org/10.7554/eLife.18554.

112 **the normal "agonist/antagonist" pairing of muscles:** Lindsay M. Biga et al., eds., "Chapter 11: The Muscular System," in *Anatomy & Physiology* (Open Oregon State: Pressbooks.com, 2018), http://library.open.oregonstate.edu/aandp/chapter/11-1-describe-the-roles-of-agonists-antagonists-and-synergists/; Janne M. Hahne et al., "Simultaneous Control of Multiple Functions of Bionic Hand Prostheses: Performance and Robustness in End Users," *Science Robotics* 3 (2018): 1–9, http://doi.org/10.1126/scirobotics.aat3630.

113 **Herr and his surgical colleagues had to redesign:** S. S. Srinivasan et al., "On Prosthetic Control: A Regenerative Agonist-Antagonist Myoneural Interface," *Science Robotics* 2, no. 6 (2017), http://doi.org/10.1126/scirobotics.aan2971.

113 **device design, computer modeling, and experimental testing:** Tyler R. Clites et al., "A Murine Model of a Novel Surgical Architecture for Proprioceptive Muscle Feedback and Its Potential Application to Control of Advanced Limb Prostheses," *Journal of Neural Engineering* 14 (2017).

114 **the first human to undergo the new procedure:** Tyler R. Clites et al., "Proprioception from a Neurally Controlled Lower-Extremity Prosthesis," *Science Translational Medicine* 10, no. 443 (2018), http://doi.org/10.1126/scitranslmed.aap8373; Gideon Gil and Matthew Orr, "Pioneering Surgery Makes a Prosthetic Foot Feel Like the Real Thing," *STAT,* May 30, 2018, http://www.statnews.com/2018/05/30/pioneering-amputation-surgery-prosthetic-foot/.

6: FEEDING THE WORLD

117 **750-square-foot "growth house":** "The Bellwether Foundation Phenotyping Facility," Donald Danforth Plant Science Center, http://www.danforthcenter.org/scientists-research/core-technologies/phenotyping.

118 **phenotypic data for their plants:** Mao Li et al., "The Persistent Homology Mathematical Framework Provides Enhanced Genotype-to-Phenotype Associations for Plant Morphology," *Plant Physiology* 177 (2018): 1382–95, http://doi.org/10.1104/pp.18.00104.

118 **phenomics refers to the full set of phenotypic information:** Robert T. Furbank and Mark Tester, "Phenomics—Technologies to Relieve the Phenotyping Bottleneck," *Trends in Plant Science* 16, no. 12 (2011): 635–44, http://doi.org/10.1016/j.tplants.2011.09.005; Daniel H. Chitwood and Christopher N. Topp, "Revealing Plant Cryptotypes: Defining Meaningful Phenotypes among Infinite Traits," *Current Opinion in Plant Biology* 24 (2015): 54–60, http://doi.org/10.1016/j.pbi.2015.01.009.

119 **genes orchestrate the development and function:** Todd P. Michael and Scott Jackson, "The First 50 Plant Genomes," *The Plant Genome* 6, no. 2 (2013): 1–7, http://doi.org/10.3835/plantgenome2013.03.0001in.

119 **by 2050, is projected to reach 9.5 billion:** United Nations Department of Economic and Social Affairs Population Division, "World Urbanization Prospects: The 2018 Revision," 2018, http://population.un.org/wup/DataQuery.

120 **double our present global crop productivity:** D. Tilman et al., "Global Food Demand and the Sustainable Intensification of Agriculture," *Proceedings of the National Academy of Sciences* 108, no. 50 (2011): 20260–64, http://doi.org/10.1073/pnas.1116437108.

120 **Evidence from archeological excavations:** M. A. Zeder, "Domestication and Early Agriculture in the Mediterranean Basin: Origins, Diffusion, and Impact," *Proceedings of the National Academy of Sciences* 105, no. 33 (2008): 11597–604, http://doi.org/10.1073/pnas.0801317105; Iosif Lazaridis et al., "Genomic Insights into the Origin of Farming in the Ancient Near East," *Nature* 536, no. 7617 (2016): 419–24, http://doi.org/10.1038/nature19310.

120 **The word *gene* dates back only to 1905:** Nils Roll-Hansen, "The Holist Tradition in Twentieth-Century Genetics. Wilhelm Johannsen's Genotype Concept," *Journal of Physiology* 592, no. 11 (2014): 2431–38, http://doi.org/10.1113/jphysiol.2014.272120; W. Johannsen, "The Genotype Conception of Heredity," *International Journal of Epidemiology* 43, no. 4 (2014): 989–1000, http://doi.org/10.1093/ije/dyu063.

121 **Mendel published his remarkable deductions:** Gregor Mendel, Versuche über Plflanzenhybriden, trans. William Bateson, *Verhandlungen des naturforschenden Vereines in Brünn, Bd. IV für das Jahr 1865*, Abhandlungen (1865): 3–47, http://www.mendelweb.org/Mendel.html; Daniel L. Hartl and Vitezslav Orel, "What Did Gregor Mendel Think He Discovered?" *Genetics* 131 (1992): 245–53, http://doi.org/10.1534/genetics.108.099762.

121 **The actual structure of DNA:** Maclyn McCarty, "Discovering Genes Are Made of DNA," *Nature* 421 (2003): 406.

123 **Michael Faraday described electromagnetic forces:** S. Chatterjee, "Michael Faraday: Discovery of Electromagnetic Induction," *Resonance* 7 (March 2002): 35–45, http://doi.org/10.1007/BF02896306.

123 **J. J. Thomson's 1897 discovery of the electron:** Joseph John Thomson, "XL. Cathode Rays," *The London, Edinburgh, and Dublin Philosophical Magazine and Journal of Science* 44, no. 269 (1897): 293–316, http://doi.org/10.1080/14786449708621070.

123 **how water crosses a cell membrane:** P. Agre et al., "Aquaporin CHIP: The Archetypal Molecular Water Channel," *American Journal of Physiology* 265 (1993): F463–76, http://doi.org/10.1085/jgp.79.5.791; Mario Parisi et al., "From Membrane Pores to Aquaporins: 50 Years Measuring Water Fluxes," *Journal of Biological Physics* 33, no. 5–6 (2007): 331–43, http://doi.org/10.1007/s10867-008-9064-5.

123 **Mendel's work was poorly understood:** Mauricio De Castro, "Johann Gregor Mendel: Paragon of Experimental Science," *Molecular Genetics and Genomic Medicine* 4, no. 1 (2016): 3–8, http://doi.org/10.1002/mgg3.199.

123 **Friedrich Miescher had isolated a substance:** Ralf Dahm, "Friedrich Miescher and the Discovery of DNA," *Developmental Biology* 278, no. 2 (2005): 274–88, http://doi.org/10.1016/j.ydbio.2004.11.028.

123 **James Watson and Francis Crick proposed the double helix:** J. D. Watson and F. H. Crick, "Molecular Structure of Nucleic Acids: A Structure for Deoxyribose Nucleic Acid," *Nature* 171, no. 4356 (1953): 737–38; Francis Crick, "Central Dogma of Molecular Biology," *Nature* 227 (1970): 561–63.

124 **established methods for transferring DNA:** R. T. Fraley et al., "Expression of Bacterial Genes in Plant Cells," *Proceedings of the National Academy of Sciences* 80, no. 15 (1983): 4803–7, http://doi.org/10.1073/pnas.80.15.4803; P. Zambryski et al., "Ti Plasmid Vector for the Introduction of DNA into Plant Cells without Alteration of Their Normal Regeneration Capacity," *EMBO Journal* 2, no. 12 (1983): 2143–50, http://doi.org/10.1002/J.1460-2075.1983.TB01715.X.

124 **genetically engineering insect-resistant tobacco:** Mark Vaeck et al., "Transgenic Plants Protected from Insect Attack," *Nature* 328, no. 6125 (1988): 33–37, http://doi.org/10.1038/328033a0.

124 **high-yield strains of corn, wheat, rice:** Elizabeth Nolan and Paulo Santos, "The Contribution of Genetic Modification to Changes in Corn Yield in the United States," *American Journal of Agricultural Economics* 94, no. 5 (2012): 1171–88, http://doi.org/10.1093/ajae/aas069; Zhi Kang Li and Fan Zhang, "Rice Breeding in the Post-Genomics Era: From Concept to Practice," *Current Opinion in Plant Biology* 16, no. 2 (2013): 261–69, http://doi.org/10.1016/j.pbi.2013.03.008.

124 **increasingly efficient fertilizers, irrigation systems:** Andrew Balmford, Rhys Green, and Ben Phalan, "Land for Food & Land for Nature?," *Daedalus* 144, no. 4 (2015): 57–75, http://doi.org/10.1162/DAED_a_00354.

124 **great surges in crop productivity:** Sun Ling Wang et al., "Agricultural Productivity Growth in the United States: Measurement, Trends and Drivers," United States Department of Agriculture Economic Research Service, 2015, http://www.ers.usda.gov/webdocs/publications/45387/53417_err189.pdf?v=42212.

124 **From the 1860s to the late 1930s:** United States Department of Agriculture National Agricultural Statistics Service, "Crop Production Historical Track Records (April 2017)," 2017, http://www.nass.usda.gov/Publications/Todays_Reports/reports/croptr17.pdf.

124 **that figure consistently reaches above 150 bushels:** United States Department of Agriculture Economic Research Service, "Corn and Other Feed Grains: Background." Last modified May 15, 2018, http://www.ers.usda.gov/topics/crops/corn-and-other-feedgrains/background/.

125 **800 million people still live in food scarcity:** Sean Sanders, ed., "Addressing Malnutrition to Improve Global Health," *Science* 346 (2014), http://doi.org/10.1126/science.346.6214.1247-d; FAO, IFAD, and WFP, *The State of Food Insecurity in the World 2014. Strengthening the enabling environment for food security and nutrition* (Rome: FAO, 2014), http://www.fao.org/3/a-i4030e.pdf.

125 **over 3 million children under the age of five:** United Nations Information Centre Canberra, "WHO Hunger Statistics," http://un.org.au/2014/05/14/who-hunger-statistics/.

125 **Norman Borlaug, one of the most important architects:** Norman E. Borlaug, "The Green Revolution Revisited and the Road Ahead," in *Nobel Prize Symposium*, 2002, http://doi.org/10.1086/451354.

125 **Plant biologists engineered the FlavrSavr:** G. Bruening and J. M. Lyons, "The Case of the FLAVR SAVR Tomato," *California Agriculture* 54, no. 4 (2000).

125 **genetic modifications that make it resistant to insect pests:** United States Department of Agriculture Economic Research Service, "Farm Practices & Management: Biotechnology Overview." Last modified January 11, 2018, http://www.ers.usda.gov/topics/farm-practices-management/biotechnology/.

126 **As recently reported by the National Academy of Sciences:** "Genetically Engineered Crops: Experiences and Prospects," The National Academies Press, 2016, http://doi.org/10.17226/23395.

126 **genes that could predispose an individual to autism or schizophrenia:** Ryan K. C. Yuen et al., "Whole Genome Sequencing Resource Identifies 18 New Candidate Genes for Autism Spectrum Disorder," *Nature Neuroscience* 20, no. 4 (2017): 602–11, http://doi.org/10.1038/nn.4524; Stephan Ripke et al., "Biological Insights from 108 Schizophrenia-Associated Genetic Loci," *Nature* 511, no. 7510 (2014): 421–27, http://doi.org/10.1038/nature13595; Aswin Sekar et al., "Schizophrenia Risk

from Complex Variation of Complement Component 4," *Nature* 530, no. 7589 (2016): 177–83, http://doi.org/10.1038/nature16549.

127 **Corn and other crops have been engineered:** Mohamed A. Ibrahim et al., "Bacillus Thuringiensis: A Genomics and Proteomics Perspective," *Bioengineered Bugs* 1, no. 1 (2010): 31–50, http://doi.org/10.4161/bbug.1.1.10519.

127 **Bt-variants of cotton and soybeans:** National Research Council of the National Academies, *Toward Sustainable Agricultural Systems in the 21st Century*, 2010, http://www.nap.edu/catalog/12832/toward-sustainable-agricultural-systems-in-the-21st-century.

127 **A field of herbicide-resistant corn:** Luca Comai, Louvminia C. Sen, and David M. Stalker, "An Altered AroA Gene Product Confers Resistance to the Herbicide Glyphosate," *Science* 221 (1983): 370–71.

128 **weed control reduces the requirement for tilling:** Jon Entine and Rebecca Randall, "GMO Sustainability Advantage? Glyphosate Spurs No-Till Farming, Preserving Soil Carbon," Genetic Literacy Project, 2017, http://geneticliteracyproject.org/2017/05/05/gmo-sustainability-advantage-glyphosate-sparks-no-till-farming-preserving-soil-carbon/.

128 **concerns about glyphosate's safety:** The National Academies Press, "Genetically Engineered Crops: Experiences and Prospects," 2016, http://doi.org/10.17226/23395.

128 **a variant known as Golden Rice:** J. Madeleine Nash, "This Rice Could Save a Million Kids a Year," *Time Magazine*, July 31, 2000, 1–7, http://content.time.com/time/magazine/article/0,9171,997586-4,00.html; Ingo Potrykus, "The 'Golden Rice' Tale," *AgBioWorld*, 2011, http://www.agbioworld.org/biotech-info/topics/goldenrice/tale.html.

128 **an insufficiency of vitamin A:** J. H. Humphrey, K. P. West, and A. Sommer, "Vitamin A Deficiency and Attributable Mortality among Under-5-Year-Olds," *Bulletin of the World Health Organization* 70, no. 2 (1992): 225–32, http://www.pubmedcentral.nih.gov/articlerender.fcgi?artid=2393289&tool=pmcentrez&rendertype=abstract.

128 **the safety and the benefits of Golden Rice:** A. Alan Moghissi, Shiqian Pei, and Yinzuo Liu, "Golden Rice: Scientific, Regulatory and Public Information Processes of a Genetically Modified Organism," *Critical Reviews in Biotechnology* 36, no. 3 (2016): 535–41, http://doi

.org/10.3109/07388551.2014.993586; Janel M. Albaugh, "Golden Rice: Effectiveness and Safety, A Literature Review," *Honors Research Projects 382*, University of Akron, 2016, http://ideaexchange.uakron.edu/honors_research_projects/382/.

128 **Australia and New Zealand approved Golden Rice:** Gary Scattergood, "Australia, New Zealand Approve Purchasing of GMO Golden Rice to Tackle Vitamin-A Deficiency in Asia," Genetic Literacy Project, 2018, http://geneticliteracyproject.org/2018/01/29/australia-new-zealand-approve-sale-gmo-golden-rice-effort-boost-fight-vitamin-deficiency-asia/.

128 **concerns about the safety and economics of genetically modified crops:** Peggy G. Lemaux, "Genetically Engineered Plants and Foods: A Scientist's Analysis of the Issues (Part I)," *Annual Review of Plant Biology* 59, no. 1 (2008): 771–812, http://doi.org/10.1146/annurev.arplant.58.032806.103840; Wilhelm Klümper and Matin Qaim, "A Meta-Analysis of the Impacts of Genetically Modified Crops," *PLoS ONE* 9, no. 11 (2014), http://doi.org/10.1371/journal.pone.0111629; Mark Lynas, "How I Got Converted to G.M.O. Food," *New York Times*, April 25, 2015, http://www.nytimes.com/2015/04/25/opinion/sunday/how-i-got-converted-to-gmo-food.html; Mitch Daniels, "Avoiding GMOs Isn't Just Anti-Science. It's Immoral," *Washington Post*, December 27, 2017, http://www.washingtonpost.com/opinions/avoiding-gmos-isnt-just-anti-science-its-immoral/2017/12/27/fc773022-ea83-11e7-b698-91d4e35920a3_story.html?noredirect=on&utm_term=.ec447407b07d; Michael Gerson, "Are You Anti-GMO? Then You're Anti-Science, Too," *Washington Post*, May 3, 2018, http://www.washingtonpost.com/opinions/are-you-anti-gmo-then-youre-anti-science-too/2018/05/03/cb42c3ba-4ef4-11e8-af46-b1d6dc0d9bfe_story.html?utm_term=.0bc14d1df5c0.

129 **complex traits that affect agricultural productivity:** Wangxia Wang, Basia Vinocur, and Arie Altman, "Plant Responses to Drought, Salinity and Extreme Temperatures: Towards Genetic Engineering for Stress Tolerance," *Planta* 218, no. 1 (2003): 1–14, http://doi.org/10.1007/s00425-003-1105-5; Huayu Sun et al., "The Bamboo Aquaporin Gene PeTIP4;1–1 Confers Drought and Salinity Tolerance in Transgenic

Arabidopsis," *Plant Cell Reports* 36, no. 4 (2017): 597–609, http://doi.org/10.1007/s00299-017-2106-3; Kathleen Greenham et al., "Temporal Network Analysis Identifies Early Physiological and Transcriptomic Indicators of Mild Drought in Brassica Rapa," *ELife* 6 (2017): 1–26, http://doi.org/10.7554/eLife.29655.

129 **This process, known as high-throughput phenotyping:** Andrade Sanchez, "Field-Based Phenomics for Plant Genetics Research," *Field Crops Research* 133 (2012): 101–12, http://doi.org/10.1080/10643389.2012.728825; J. L. Araus and J. E. Cairns, "Field High-Throughput Phenotyping: The New Crop Breeding Frontier," *Trends in Plant Science* 19, no. 1 (2014): 52–61, http://doi.org/10.1016/j.tplants.2013.09.008; Noah Fahlgren et al., "A Versatile Phenotyping System and Analytics Platform Reveals Diverse Temporal Responses to Water Availability in Setaria," *Molecular Plant* 8, no. 10 (2015): 1520–35, http://doi.org/10.1016/j.molp.2015.06.005; Malia A. Gehan and Elizabeth A. Kellogg, "High-Throughput Phenotyping," *American Journal of Botany* 104, no. 4 (2017): 505–8, http://doi.org/10.3732/ajb.1700044; Jordan R. Ubbens and Ian Stavness, "Deep Plant Phenomics: A Deep Learning Platform for Complex Plant Phenotyping Tasks," *Frontiers in Plant Science* 8 (July 2017), http://doi.org/10.3389/fpls.2017.01190.

130 **Wild soybeans from China can contribute nematode resistance:** Xue Zhao et al., "Loci and Candidate Genes Conferring Resistance to Soybean Cyst Nematode HG Type 2.5.7," *BMC Genomics* 18, no. 1 (2017): 1–10, http://doi.org/10.1186/s12864-017-3843-y.

130 **mixing close to 40,000 different genes:** Jeremy Schmutz et al., "Genome Sequence of the Palaeopolyploid Soybean," *Nature* 463, no. 7278 (2010): 178–83, http://doi.org/10.1038/nature08670.

131 **Barbara McClintock, the great twentieth-century corn geneticist:** Barbara McClintock, "The Origin and Behavior of Mutable Loci in Maize," *Proceedings of the National Academy of Sciences* 36 (1950): 344–55; Barbara McClintock, "The Significance of Responses to the Genome to Challenge: Nobel Lecture," 1983, http://www.nobelprize.org/nobel_prizes/medicine/laureates/1983/mcclintock-lecture.html.

131 **turning to drones and satellite imagery:** John Wihbey, "Agricultural Drones May Change the Way We Farm," *Boston Globe*, August

23, 2015, http://www.bostonglobe.com/ideas/2015/08/22/agricultural-drones-change-way-farm/WTpOWMV9j4C7kchvbmPr4J/story.html; Steve Curwood and Nikhil Vadhavkar, "Drones Are the Future of Agriculture," *Living on Earth*, August 5, 2016, http://www.loe.org/shows/segments.html?programID=16-P13-00032&segmentID=5; G. Lobet, "Image Analysis in Plant Sciences: Publish Then Perish," *Trends in Plant Science* 22 (2017): 1–8, http://doi.org/10.1016/j.tplants.2017.05.002.

133 **"to improve the human condition through plant science":** "Improving the Human Condition through Plant Science," Donald Danforth Plant Science Center: Roots & Shoots Blog. Last modified January 6, 2015, http://www.danforthcenter.org/news-media/roots-shoots-blog/blog-item/improving-the-human-condition-through-plant-science.

133 **I was met at the Danforth Center:** Elizabeth Kellogg in discussion with the author, October 2017.

133 **studying cereals and their relatives in the grass family:** Elizabeth Kellogg, "Relationships of Cereal Crops and Other Grasses," *Proceedings of the National Academy of Sciences* 95, no. 5 (1998): 2005–10, http://doi.org/10.1073/pnas.95.5.2005.

133 **Cereal crops have anchored human nutrition:** Carl Zimmer, "Where Did the First Farmers Live? Looking for Answers in DNA," *New York Times*, October 18, 2016, http://www.nytimes.com/2016/10/18/science/ancient-farmers-archaeology-dna.html.

134 **Kellogg took me to meet Dr. Jim Carrington:** Jim Carrington in discussion with the author, October 2017.

135 **using RNA to inhibit gene expression:** Jia He et al., "Threshold-Dependent Repression of SPL Gene Expression by MiR156/MiR157 Controls Vegetative Phase Change in Arabidopsis Thaliana," *PLoS Genetics* 14, no. 4 (2018): 1–28, http://doi.org/10.1371/journal.pgen.1007337.

136 **adapts X-ray technologies used to detect metal fatigue:** A. Tabb, K. E. Duncan, and C. N. Topp, "Segmenting Root Systems in X-Ray Computed Tomography Images Using Level Sets," 2018 IEEE Winter Conference on Applications of Computer Vision (WACV), Lake Tahoe, NV/CA, 586–595, http://doi.org/10.1109/wacv.2018.00070.

136 **by applying exogenous nitrogen in fertilizer:** National Research

Council of the National Academies, *Toward Sustainable Agricultural Systems in the 21st Century*, 2010, http://www.nap.edu/catalog/12832/toward-sustainable-agricultural-systems-in-the-21st-century.

136 **I headed off with two Danforth scientists:** Becky Bart and Nigel Taylor in discussion with the author, October 2017.

136 **the Danforth's greenhouse complex of forty-three experimental stations:** Donald Danforth Plant Science Center, "Campus: Donald Danforth Plant Science Center Facility," http://www.danforthcenter.org/about/campus.

137 **drought poses perhaps the most significant challenge:** Shujun Yang et al., "Narrowing down the Targets: Towards Successful Genetic Engineering of Drought-Tolerant Crops," *Molecular Plant* 3, no. 3 (2010): 469–90, http://doi.org/10.1093/mp/ssq016; Andrew Marshall, "Drought-Tolerant Varieties Begin Global March," *Nature Biotechnology* 32, no. 4 (2014): 308, http://doi.org/10.1038/nbt.2875; Mark Cooper et al., "Breeding Drought-Tolerant Maize Hybrids for the US Corn-Belt: Discovery to Product," *Journal of Experimental Botany* 65, no. 21 (2014): 6191–94, http://doi.org/10.1093/jxb/eru064.

138 **improvement compared to doing the work manually:** Elizabeth Kellogg in discussion with the author, April 2018.

138 **software for this purpose, called PlantCV:** Malia A. Gehan et al., "PlantCV v2: Image Analysis Software for High-Throughput Plant Phenotyping," *PeerJ* 5 (2017): e4088, http://doi.org/10.7717/peerj.4088.

138 **Todd Mockler, another of Danforth Center's Distinguished Investigators:** Todd Mockler in discussion with the author, October 2017.

138 **a field-testing facility in Maricopa, Arizona:** United States Department of Agriculture, "Plant Physiology and Genetics Research: Maricopa, AZ," http://www.ars.usda.gov/pacific-west-area/maricopa-arizona/us-arid-land-agricultural-research-center/plant-physiology-and-genetics-research/.

138 **The project, TERRA-REF:** Noah Fahlgren, Malia A. Gehan, and Ivan Baxter, "Lights, Camera, Action: High-Throughput Plant Phenotyping Is Ready for a Close-Up," *Current Opinion in Plant Biology* 24 (2015): 93–99, http://doi.org/10.1016/j.pbi.2015.02.006; Nadia Shakoor, Scott Lee, and Todd C. Mockler, "High-Throughput Phenotyping to Accel-

erate Crop Breeding and Monitoring of Diseases in the Field," *Current Opinion in Plant Biology* 38 (2017): 184–92, http://doi.org/10.1016/j.pbi.2017.05.006.

139 **advanced computational methods, including machine learning:** Arti Singh et al., "Machine Learning for High-Throughput Stress Phenotyping in Plants," *Trends in Plant Science* 21, no. 2 (2016): 110–24, http://doi.org/10.1016/j.tplants.2015.10.015; Sotirios A. Tsaftaris, Massimo Minervini, and Hanno Scharr, "Machine Learning for Plant Phenotyping Needs Image Processing," *Trends in Plant Science* 21, no. 12 (2016): 989–91, http://doi.org/10.1016/j.tplants.2016.10.002; Pouria Sadeghi-Tehran et al., "Multi-Feature Machine Learning Model for Automatic Segmentation of Green Fractional Vegetation Cover for High-Throughput Field Phenotyping," *Plant Methods* 13, no. 1 (2017): 1–16, http://doi.org/10.1186/s13007-017-0253-8.

139 **other kinds of artificial intelligence:** Frank Vinluan, "A.I.'s Role in Agriculture Comes into Focus with Imaging Analysis," *Xconomy*, May 2, 2017, http://www.xconomy.com/raleigh-durham/2017/05/02/a-i-s-role-in-agriculture-comes-into-focus-with-imaging-analysis/.

139 **a recent report published by the US Department of Agriculture:** United States Department of Agriculture, Economic Research Service using data from the National Agricultural Statistics Service, "June Agricultural Survey." Last updated July 12, 2017, http://www.ers.usda.gov/data-products/adoption-of-genetically-engineered-crops-in-the-us.aspx.

139 **more than 90 percent of the acres farmed in the United States:** "Adoption of Genetically Engineered Cotton in the United States, by Trait, 2000–17," http://www.ers.usda.gov/webdocs/charts/56323/biotechcotton.png?v=42565; "Adoption of Genetically Engineered Corn in the United States, by Trait, 2000–17," http://www.ers.usda.gov/webdocs/charts/55237/biotechcorn.png?v=42565.

140 **staple foods for more than half a billion people:** FAO and IFAD, *The World Cassava Economy* (Rome: International Fund for Agricultural Development and Food and Agriculture Organization of the United Nations, 2000), http://www.fao.org/docrep/009/x4007e/X4007E00.htm#TOC.

140 **Cassava brown streak disease (CBSD), an insect-transmitted viral disease:** "VIRCA Plus: Virus-Resistant and Nutritionally-Enhanced Cassava for Africa," Donald Danforth Plant Science Center. Last modified November 2017, http://www.danforthcenter.org/scientists-research/research-institutes/institute-for-international-crop-improvement/crop-improvement-projects/virca-plus.

140 **the plant is genetically modified:** Donald Danforth Plant Science Center, "New Cassava Potential—VIRCA—Fact Sheet," http://www.danforthcenter.org/scientists-research/research-institutes/institute-for-international-crop-improvement/crop-improvement-projects/virca.

141 **farmers in East Africa may have access:** Nigel Taylor in discussion with the author, April 2018.

141 **using X-ray technology to image roots:** A. Bucksch et al., "Image-Based High-Throughput Field Phenotyping of Crop Roots," *Plant Physiology* 166, no. 2 (2014): 470–86, http://doi.org/10.1104/pp.114.243519.

142 **Cassava storage roots are an excellent source of calories:** Donald Danforth Plant Science Center, "VIRCA Plus: Virus-Resistant and Nutritionally-Enhanced Cassava for Africa." Last modified November 2017, http://www.danforthcenter.org/scientists-research/research-institutes/institute-for-international-crop-improvement/crop-improvement-projects/virca-plus.

143 **more than 9.5 billion people all over the planet:** United Nations Department of Economic and Social Affairs Population Division, "World Urbanization Prospects: The 2018 Revision," 2018, http://population.un.org/wup/DataQuery.

7: CHEATING MALTHUS, ONCE AGAIN

145 **Karl Taylor Compton wrote a delightful article:** Karl T. Compton, "The Electron: Its Intellectual and Social Significance," *Nature* 139, no. 3510 (1937): 229–40.

145 **Thomson won the Nobel Prize in 1906:** J. J. Thomson, "Carriers of Negative Electricity," *Nobel Lecture*, 1906, https://www.nobelprize.org/nobel_prizes/physics/laureates/1906/thomson-lecture.html.

146 **component parts: namely, quarks and gluons:** Michael Riordan, "The Discovery of Quarks," *Science* 256 (1992): 1287–93; John Ellis, "The Discovery of the Gluon," *ArXiv*, 2014, http://arxiv.org/pdf/1409.4232.pdf.

146 **the father of nuclear physics:** E. Rutherford, "LXXIX. The Scattering of α and β Particles by Matter and the Structure of the Atom," *Philosophical Magazine Series 6* 21, no. 125 (1911): 669–88, http://doi.org/10.1080/14786440508637080.

146 **"anyone who expects a source of power":** B. Cameron Reed, "A Compendium of Striking Manhattan Project Quotes," *History of Physics Newsletter* 13, no. 3 (2016): 8, http://doi.org/10.1016/j.chembiol.2011.05.005.

147 **a nuclear power plant at the Idaho National Laboratory:** "Experimental Breeder Reactor-I," Idaho National Laboratory. Last modified February 8, 2012, http://www4vip.inl.gov/research/experimental-breeder-reactor-1/d/experimental-breeder-reactor-1.pdf.

147 **British Chancellor of the Exchequer, William Gladstone:** William Edward Hartpole Lecky, *Democracy and Liberty* (New York: Longmans, Green, and Co., 1899), http://oll.libertyfund.org/titles/1813.

148 **this convergence—Convergence 2.0:** P. Sharp, T. Jacks, and S. Hockfield, "Capitalizing on Convergence for Health Care," *Science* 352, no. 6293 (2016): 1522–23, http://doi.org/10.1126/science.aag2350; Phillip Sharp and Susan Hockfield, "Convergence: The Future of Health," *Science* 355, no. 6325 (2017): 589, http://doi.org/10.1126/science.aam8563.

148 **Peter Holme Jensen, told me:** Peter Holme Jensen in discussion with the author, September 2017.

148 **turning methane into ethylene:** Alexander H. Tullo, "Ethylene from Methane: Researchers Take a New Look at an Old Problem," *Chemical and Engineering News* 89, no. 3 (2011): 20–21.

149 **reverse climate change by capturing carbon dioxide:** Ik Dong Choi, Jae Won Lee, and Yong Tae Kang, "CO_2 Capture/Separation Control by SiO_2 Nanoparticles and Surfactants," *Separation Science and Technology* 50, no. 5 (2015): 772–80, http://doi.org/10.1080/01496395.2014.965257; Alison E. Berman, "How Nanotech

Will Lead to a Better Future for Us All," 2016, http://singularityhub.com/2016/08/12/how-nanotech-will-lead-to-a-better-future-for-us-all/#sm.000f3rwrf13l3epptd91mp6z4gw9y.

149 **render just about any surface self-cleaning:** Ivan P. Parkin and Robert G. Palgrave, "Self-Cleaning Coatings," *Journal of Materials Chemistry* 15, no. 17 (2005): 1689–95, http://doi.org/10.1039/b412803f.

149 **Unemployment stood at over 14 percent:** Roosevelt Institute, "Roosevelt Recession." Last modified August 19, 2010, http://rooseveltinstitute.org/roosevelt-recession/.

151 **the gloomy future that in 1798 Thomas Malthus described:** Thomas Robert Malthus, "An Essay on the Principle of Population as It Affects the Future Improvement of Society," 1798.

152 **"the lessons learned in the war-time":** Vannevar Bush, "Science: The Endless Frontier, A Report to the President by Vannevar Bush, Director of the Office of Scientific Research and Development," Washington, DC, 1945, http://www.nsf.gov/od/lpa/nsf50/vbush1945.htm.

152 **the level of federal investment in R&D:** Mark Boroush, "U.S. R&D Increased by $20 Billion in 2015, to $495 Billion; Estimates for 2016 Indicate a Rise to $510 Billion," *NCSES InfoBrief*, 2017, http://www.nsf.gov/statistics/2018/nsf18306/nsf18306.pdf.

152 **government has scaled back its investment:** Stephen A. Merrill, "Righting the Research Imbalance," 2018.

153 **private philanthropic funds for basic science research:** The Science Philanthropy Alliance, "2016 Survey of Private Funding for Basic Research Summary Report," 2016, http://www.sciencephilanthropyalliance.org/wp-content/uploads/2017/02/Survey-of-Private-Funding-for-Basic-Research-Summary-021317.pdf.

153 **The American Association for the Advancement of Science reported:** AAAS, "R&D Budget and Policy Program: Research by Science and Engineering Discipline." Last modified September 2017, http://www.aaas.org/page/research-science-and-engineering-discipline.

154 **Between 1995 and 2015 many nations:** National Science Board, *Science and Engineering Indicators 2018, NSB-2018-1* (Alexandria, VA: National Science Foundation, 2018), http://www.nsf.gov/statistics/indicators/.

154 **the National Institutes of Health, with an annual budget:** National

Institutes of Health, "Impact of NIH Research." Last modified May 1, 2018, http://www.nih.gov/about-nih/what-we-do/impact-nih-research/our-society.

155 **a rise in the average life expectancy for Americans:** The World Bank Group, "Life Expectancy at Birth, Total (Years)." Last modified 2017, http://data.worldbank.org/indicator/SP.DYN.LE00.IN?end=2015&locations=US&start=1960.

155 **the Human Genome Project, launched in 1990:** Francis S. Collins et al., "New Goals for the U.S. Human Genome Project: 1998–2003," *Science* 282, no. 5389 (1998): 682–89, http://doi.org/10.1126/science.282.5389.682.

155 **the first maps of the fly, mouse, and human genomes:** E. S. Lander et al., "Initial Sequencing and Analysis of the Human Genome," *Nature* 409, no. 6822 (2001): 860–921, http://doi.org/10.1038/35057062; J. C. Venter et al., "The Sequence of the Human Genome," *Science* 291, no. 5507 (2001): 1304–51, http://doi.org/10.1126/science.1058040; R. H. Waterston et al., "Initial Sequencing and Comparative Analysis of the Mouse Genome," *Nature* 420, no. 6915 (2002): 520–62, http://doi.org/10.1038/nature01262\rnature01262 [pii]. Mark D. Adams et al., "The Genome Sequence of Drosophila Melanogaster," *Science* 287 (2017): 2185–96.

155 **candidate genes for cancer, diabetes, schizophrenia:** Yoshio Miki et al., "Strong Candidate for the Breast and Ovarian Cancer Susceptibility Gene BRCA1," *Science* 266, no. 5182 (1994): 66–71, http://doi.org/10.1126/science.7545954; Décio L. Eizirik et al., "The Human Pancreatic Islet Transcriptome: Expression of Candidate Genes for Type 1 Diabetes and the Impact of Pro-Inflammatory Cytokines," *PLoS Genetics* 8, no. 3 (2012), http://doi.org/10.1371/journal.pgen.1002552; Tiffany A. Greenwood et al., "Association Analysis of 94 Candidate Genes and Schizophrenia-Related Endophenotypes," *PLoS ONE* 7, no. 1 (2012), http://doi.org/10.1371/journal.pone.0029630.

155 **sequencing a human genome cost:** National Human Genome Research Institute, "DNA Sequencing Costs: Data." Last modified April 25, 2018, http://www.genome.gov/sequencingcostsdata/.

156 **the National Nanotechnology Initiative (NNI):** National Academies of Sciences, Engineering, and Medicine, *Triennial Review of the*

National Nanotechnology Initiative (Washington, DC: National Academies Press, 2016), http://doi:10.17226/23603.

156 **Brain Research through Advancing Innovative Neurotechnologies Initiative (BRAIN):** Cornelia I. Bargmann and William T. Newsome, "The Brain Research Through Advancing Innovative Neurotechnologies (BRAIN) Initiative and Neurology," *JAMA Neurology* 71, no. 6 (2014): 675–76, http://doi.org/10.1001/jamaneurol.2014.411.Conflict.

157 **Paula Hammond, a chemical engineer:** Sarah L. Clark and Paula T. Hammond, "Engineering the Microfabrication of Layer-by-Layer Thin Films," *Advanced Materials* 10, no. 18 (1998): 1515–19.

158 **Michael Yaffe, a physician and molecular biologist:** Andrew E. H. Elia, Lewis C. Cantley, and Michael B. Yaffe, "Proteomic Screen Finds PSer/PThr-Binding Domain Localizing Plk1 to Mitotic Substrates," *Science* 299, no. 5610 (2003): 1228–31, http://doi.org/10.1126/science.1079079.

158 **amplify the effectiveness of chemotherapy:** Erik C. Dreaden et al., "Tumor-Targeted Synergistic Blockade of MAPK and PI3K from a Layer-by-Layer Nanoparticle," *Clinical Cancer Research* 21, no. 5 (2015): 4410–20, http://doi.org/10.1158/1078-0432.CCR-15-0013.

160 **passage of the Bayh-Dole Act:** David C. Mowery and Bhaven N. Sampat, "The Bayh-Dole Act of 1980 and University-Industry Technology Transfer: A Model for Other OECD Governments?" *Journal of Technology Transfer* 30, no. 1/2 (2005): 115–27, http://doi.org/10.1007/0-387-25022-0_18.

160 **the number of patents issued to US academic institutions:** Ampere A. Tseng and Miroslav Raudensky, "Assessments of Technology Transfer Activities of US Universities and Associated Impact of Bayh-Dole Act," *Scientometrics* 101, no. 3 (2014): 1851–69, http://doi.org/10.1007/s11192-014-1404-6; National Science Board, *Science and Engineering Indicators 2018, NSB-2018-1* (Alexandria, VA: National Science Foundation, 2018), http://www.nsf.gov/statistics/indicators/.

160 **the top few in the US Patent and Trade Office's annual report:** U.S. Patent and Trademark Office Patent Technology Monitoring Team, U.S. Colleges and Universities—Utility Patent Grants, Calendar Years 1969–2012, 2012, http://www.uspto.gov/web/offices/ac/ido/oeip/taf/univ/univ_toc.htm.

163 **Hélix-Nielsen explained that the major investors:** Claus Hélix-Nielsen in discussion with the author, September 2017.

163 **The government, Fink suggests:** Larry Fink, "Larry Fink's Annual Letter to CEOs: A Sense of Purpose," Blackrock, 2017, http://www.blackrock.com/corporate/investor-relations/larry-fink-ceo-letter.

163 **successful American companies with first- or second-generation immigrants:** Center for American Entrepreneurship, "Immigrant Founders of the 2017 Fortune 500," accessed June 18, 2018, http://startupsusa.org/fortune500/.

163 **nearly half of the founders of the Fortune 500 companies:** Leigh Buchanan, "Study: Nearly Half the Founders of America's Biggest Companies Are First- or Second-Generation Immigrants," *Inc.*, December 5, 2017, http://www.inc.com/leigh-buchanan/fortune-500-immigrant-founders.html.

163 **a third of all new graduate students:** National Science Board, "Higher Education in Science and Engineering," in *Science and Engineering Indicators 2018, 2.1-109* (Arlington, VA: National Science Foundation, 2018), http://www.nsf.gov/statistics/seind12/.

164 **International student enrollment in graduate programs:** Nick Anderson, "Report Finds Fewer New International Students on U.S. College Campuses," *Washington Post*, November 12, 2017, http://www.washingtonpost.com/local/education/report-finds-fewer-new-international-students-on-us-college-campuses/2017/11/12/5933fe02-c61d-11e7-aae0-cb18a8c29c65_story.html.

164 **the number of new international students enrolling:** Hironao Okahana and Enyu Zhou, *International Graduate Applications and Enrollment: Fall 2017* (Washington, DC: Council of Graduate Schools, 2018).

164 **declined for the first time since the 2001 terrorist attacks:** Bianca Quilantan, "International Grad Students' Interest in American Higher Ed Marks First Decline in 14 Years," *Chronicle of Higher Education*, January 30, 2018, http://www.chronicle.com/article/International-Grad-Students-/242377.

164 **save us from the impending crises:** Ted Nordhaus, "The Earth's Carrying Capacity for Human Life Is Not Fixed," *Aeon*, July 5, 2018,

http://aeon.co/ideas/the-earths-carrying-capacity-for-human-life-is-not-fixed.

164 **a global population of more than 9.5 billion:** United Nations Department of Economic and Social Affairs Population Division, "World Urbanization Prospects: The 2018 Revision," 2018, http://population.un.org/wup/DataQuery.

166 **Nobel Prize–winning biologist Phillip Sharp:** Phillip A. Sharp, "Split Genes and RNA Splicing," *Nobel Lectures*, 1993, http://www.nobelprize.org/nobel_prizes/medicine/laureates/1993/sharp-lecture.pdf.

INDEX

Note: Page numbers in italics indicate figures.